Use R!

Series Editors:
Robert Gentleman   Kurt Hornik   Giovanni Parmigiani

More information about this series at http://www.springer.com/series/6991

Use R!

Albert: Bayesian Computation with R (2nd ed. 2009)

Bivand/Pebesma/Gómez-Rubio: Applied Spatial Data Analysis with R (2nd ed. 2013)

Cook/Swayne: Interactive and Dynamic Graphics for Data Analysis: With R and GGobi

Hahne/Huber/Gentleman/Falcon: Bioconductor Case Studies

Paradis: Analysis of Phylogenetics and Evolution with R (2nd ed. 2012)

Pfaff: Analysis of Integrated and Cointegrated Time Series with R (2nd ed. 2008)

Sarkar: Lattice: Multivariate Data Visualization with R

Spector: Data Manipulation with R

Douglas A. Luke

A User's Guide to Network Analysis in R

 Springer

Douglas A. Luke
Center for Public Health Systems Science
George Warren Brown School
  of Social Work
Washington University
St. Louis, MO, USA

ISSN 2197-5736   ISSN 2197-5744   (electronic)
Use R!
ISBN 978-3-319-23882-1   ISBN 978-3-319-23883-8   (eBook)
DOI 10.1007/978-3-319-23883-8

Library of Congress Control Number: 2015955739

Springer Cham Heidelberg New York Dordrecht London
© Springer International Publishing Switzerland 2015

Printed on acid-free paper

Springer International Publishing AG Switzerland is part of Springer Science+Business Media (www.springer.com)

To my most important social network—Sue, Alina, and Andrew

Preface

In early 2000, Stephen Hawking said that "...the next century will be the century of complexity." If his prediction is true, the implication is that we will need new scientific theories, data collection methods, and analytic techniques that are appropriate for the study of complex systems and behavior. Network science is one such approach that views the world through a network lens, where physical and social systems are made up of heterogeneous actors who are connected to one another through different types of relational ties. Network analysis is the set of analytic tools used to study these types of systems. Over the past several decades network analysis has become an increasingly important part of the analytic toolbox for social, health, and physical scientists.

Until recently, network analysis required specialized software, both for network data management and analyses. However, starting around 2000, network analytic tools became available in the R statistical programming environment. This not only made network analytic techniques more visible to the broader statistical community but also provided the breadth and power of R's data management, graphic visualization, and general statistical modeling capabilities to the network analyst community.

As the title suggests, this book is a user's guide to network analysis in R. It provides a practical hands-on tour of the major network analytic tasks that can currently be done in R. The book concentrates on four primary tasks that a network analyst typically concerns herself with: network data management, network visualization, network description, and network modeling. The book includes all the R code that is used in the network analysis examples. It also comes with a set of network datasets that are used throughout the book. (See Chap. 1 for more details on the structure of the book, as well as instructions on how to obtain the network data.) The book is written for anybody who has an interest in doing network analysis in R. It can be used as a secondary text in a network science or analysis class or can simply serve as a reference for network techniques in R.

This book would not exist without the help, support, guidance, and mentoring I have received over the last 30 years from my own personal and professional social networks. In the mid-1980s I took a graduate network analysis class from Stan Wasserman at the University of Illinois in Champaign. I remember being excited about this new way to analyze data, but thought that I was not likely to ever use it in my career. However, my colleagues in psychology and public health encouraged me in my early work exploring how network analysis could answer important research and evaluation questions. These include Julian Rappaport, Ed Seidman, Bruce Rapkin, Kurt Ribisl, Sharon Homan, Ross Brownson, and Matt Kreuter. Whether they know it or not, I have been inspired and encouraged by an amazing group of network and systems scientists, including Tom Valente, Steve Borgatti, Martina Morris, Tom Snijders, Scott Leischow, Patty Mabry, Stephen Marcus, and Ross Hammond. My best network ideas have come from my friends and colleagues at the Center for Public Health Systems Science, particularly Bobbi Carothers, Amar Dhand, Chris Robichaux, and Nancy Mueller. I am especially grateful to the students in my network analysis classes and workshops over the years; they have not only improved this book, but they have improved my thinking about network analysis. A very special thank you to Jenine Harris. Jenine was my first doctoral student, now I am inspired by the rigor and elegance of her own work in network science. I would also like to thank the Centers for Disease Control and Prevention, the National Institutes of Health, and the Missouri Foundation for Health for providing research and evaluation support that allowed me to develop and refine my approach to network analysis. Finally, my deepest thanks go to my family. They gave me specific suggestions about the content, provided me space and time to work hard on this book (including a crucial Father's Day gift), and cheered me on when I most needed it. Thank you, Sue, Ali, and Andrew.

St. Louis, MO, USA   Douglas A. Luke
July, 2015

Contents

Chapter 1
Introducing Network Analysis in R

> *Begin at the beginning,* the King said, very gravely, *and go on till you come to the end: then stop. (*Lewis Carroll, *Alice in Wonderland)*

1.1 What Are Networks?

This book is a user's guide for conducting network analysis in the R statistical programming language. Networks are all around us. Humans naturally organize themselves in networked systems. Our families and friends form personal social networks around each of us. Neighborhoods and communities organize themselves in networked coalitions to advocate for change. Businesses work with (and against) each other in complex, interlocking networks of trade and financial partnerships. Public health is advanced through partnerships and coalitions of governmental and NGO organizations (Luke and Harris 2007). Nations are connected to one another through systems of migration, trade, and treaty obligations.

Moreover, non-human networks exist almost anywhere you look. Our genes and proteins interact with one another through complex biological networks. The human brain is now viewed as a complex network, or 'connectome' (Sporns 2012). Similarly, human diseases and their underlying genetic roots are connected as a 'diseasome' (Barabási 2007). Animal species interact in many complex ways, one of which is a networked food-web that describes interactions in 'who-eats-whom' relationships. Information itself is networked. Our legal system is built on an interconnecting network of prior legal decisions and precedents. Social and scientific progress is driven by a diffusion of innovation process by which information is disseminated across connected social systems, whether they are Iowa corn farmers (Rogers 2003) or public health scientists (Harris and Luke 2009). It appears that one of the ways the universe is organized is with networks.

So what is a network? Figures 1.1 and 1.2 present two examples of important and interesting social networks. Figure 1.1 presents the contact network of the 19 9–11 hijackers, based on the work of Valdis Krebs (2002). Every social network is made up of a set of actors (also called nodes) that are connected to one another via some type of social relationship (also called a tie). In the figure, nodes are the circles and the ties are the lines connecting some of the nodes. The network shows

© Springer International Publishing Switzerland 2015
D.A. Luke, *A User's Guide to Network Analysis in R*, Use R!,
DOI 10.1007/978-3-319-23883-8_1

us that the hijackers had some contact with one another before September 11th, but the network is not very densely connected and there appears to be no prominent network member who is connected to all or even most of the other hijackers.

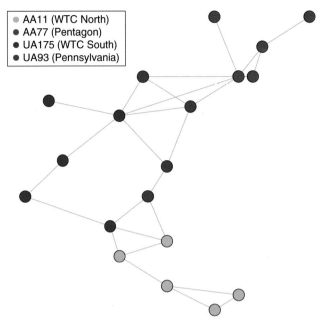

Fig. 1.1 Network of 9–11 hijackers

The second example in Fig. 1.2 is from a very different sort of social network. Here the nodes are members of the 2010 Netherlands FIFA World Cup team, who went on to lose in the final to Spain. The ties represent passes between the different players during the World Cup matches. The arrows show the directional pattern of the passes. We can see that the goalkeeper passed primarily to the defenders, and the forwards received passes primarily from the midfielders (except for #6, who appears to have a different passing pattern than the other two forwards).

These two examples may appear to have little in common. However, they both share a fundamental characteristic common to all social networks. The social patterns that are displayed in the network figures are not random. They reflect underlying social processes that can be explored using network science theories and methods. The terrorist network has no prominent leader and is not tightly interconnected because it makes the network harder to detect or disrupt. The pattern of passing ties in the soccer network reflects the assigned positions of the players, the rules of the game, and the strategies of the coach. The network analysis does not 'know' about any of those rules or strategies. Yet, network analysis can be used to reveal these patterns that reflect the underlying rules and regularities.

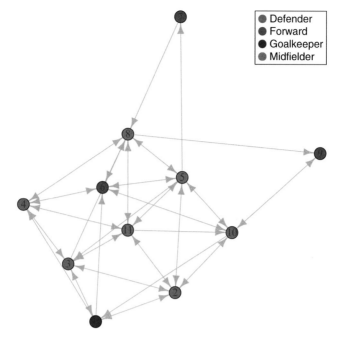

Fig. 1.2 Network of Netherlands 2010 World Cup soccer team

1.2 What Is Network Analysis?

Network science is a broad approach to research and scholarship that uses a relational lens to study and understand biological, physical, social, and informational systems. The primary tool for network scientists is network analysis, which is a set of methods that are used to (1) visualize networks, (2) describe specific characteristics of overall network structure as well as details about the individual nodes, ties, and subgroups within the networks, and (3) build mathematical and statistical models of network structures and dynamics. Because the core question of network science is about relationships, most of the methods used in network analysis are quite distinct from the more traditional statistical tools used by social and health scientists.

Network analysis as a distinct scientific enterprise with its own theories and methods grew out of developments in many other disciplines, particularly graph theory and topology in mathematics, the study of kinship systems in anthropology, and social groups and process from sociology and psychology. Although network analysis was not invented by one person at a specific place and time, the initial development of what we now recognize as modern network analysis can be traced back to the work of Jacob Moreno in the 1930s. He defined the study of social relations as sociometry, and founded the journal *Sociometry* that would publish the early studies in this area. He also invented the sociogram, which was a visual way to display network structures. The first published sociogram appeared in the New York Times in 1933, and it was a network diagram of the friendship ties among a 4th grade class. (These data are available as part of the network dataset package that accompanies this book, see Sect. 1.4.3 below.)

The theories and methods of network analysis were developed throughout the rest of the twentieth century, with important contributions from sociology, psychology, political science, business, public health, and computer science. Network science as an empirical practice was propelled by the development of a number of network specific software tools and packages, including UCINet, STRUCTURE, Negopy, and Pajek. The interest in network science has exploded in the last 20–30 years, driven by at least three different factors. First, mathematicians, physicists, and other researchers developed a number of influential theories of network structure and formation that brought attention and energy to network science (see Chap. 10 for some discussion of these theories). Second, advances in computational power and speed allowed network methods to be applied to large and very large networks, such as the internet, the population of the planet, or the human brain. Finally, advances in statistical network theory allowed analysts for the first time to move beyond simple network description to be able to build and test statistical models of network structures and processes (see Chaps. 11 and 12).

1.3 Five Good Reasons to Do Network Analysis in R

As the title suggests, this book is designed as a general guide for how to do network analysis in the R statistical language and environment. Why is R an ideal platform for developing and conducting network analyses? There are at least five good reasons.

1.3.1 Scope of R

The R statistical programming language and environment comprise a vast integrated system of thousands of packages and functions that allow it to handle innumerable data management, analysis, or visualization tasks. The R system includes a number of packages that are designed to accomplish specific network analytic tasks. However, by performing these network tasks within the R environment, the analyst can take advantage of any of the other capabilities of R. Most other network analysis programs (e.g., Pajek, UCINet, Gephi) are stand-alone packages, and thus do not have the advantages of working within an integrated statistical programming environment.

1.3.2 Free and Open Nature of R

One of the important reasons for R's popularity and success is its free and open nature. This is formally ensured via the GNU General Public License (GPL) that R-code is released under. More informally, there is a vast R user and developer community which is continually working to enhance and improve R base code and the thousands of R packages that can be freely accessed. The social network capabilities of R described in this book have, in fact, been developed by the R user community. This open nature of R facilitates faster (and arguably, cleaner and more powerful) development and dissemination of new statistical and data analytic techniques, such as these network analytic tools.

1.3.3 Data and Project Management Capabilities of R

Although there are many good network analysis programs available which can handle a wide variety of network descriptive statistics and visualization tasks, no other network package has the same power to handle often complex data and project management tasks for larger-scale network analyses compared to R. First, as suggested above, network analysis in R can take advantage of the powerful data management, cleaning, import and export capabilities of base R. As described in Chap. 3, network analysis often starts by importing and transforming data from other sources into a form that can be analyzed by network tools. All network packages have some data management capabilities, but no other program can match R's breadth and depth.

Second, when conducting sophisticated scientific or commercial network analyses, it is important to have the right project management tools to facilitate code storage and retrieval, managing analysis outputs such as statistical results and information graphics, and producing reports for internal and external audiences. Traditional statistical analysis platforms such as SAS and SPSS have these sorts of tools, but most network programs do not. By pairing R up with an integrated development environment (IDE) such as RStudio (http://rstudio.org/) and taking advantage of packages such as knitr and shiny, the user has the ability to manage any type of complex network project. In fact, the development and availability of these tools has been one of the driving forces of the *reproducible research* movement (Gentleman and Lang 2007), which emphasizes the importance of combining data, code, results, and documentation in permanent and shareable forms. As one example of the power of the reproducible research tools accessible in R is this book, which was created entirely in RStudio.

1.3.4 Breadth of Network Packages in R

The primary reason R is ideal for network analysis is the breadth of packages that are currently available to manage network data and conduct network visualization, network description, and network modeling. There are dozens of network-related packages, and more are being created all the time. R network data can be managed and stored in R native objects by the `network` and `igraph` packages, and the data can be exchanged between formats with the `intergraph` package. Basic network analysis and visualization can be handled with the `sna` package contained within the much broader `statnet` suite of network packages, as well as within `igraph`. More sophisticated network modeling can be handled by `ergm` and its associated libraries, and dynamic actor-based network models are produced by `RSiena`. Free-standing network analysis programs have many strengths (e.g., the visualization capabilities of Gephi), but no single program matches the combined power of the social network analysis packages contained in R.

1.3.5 Strength of Network Modeling in R

Finally, the particular network modeling strengths of R should be mentioned. R is the only generally available software package that includes comprehensive facilities to do stochastic network modeling (e.g., exponential random graph models), dynamic actor-based network models that allow study of how networks change over time, and other network simulation procedures.

1.4 Scope of Book and Resources

1.4.1 Scope

As the title suggests, the goal of this book is to provide a hands-on, practical guide to doing network analysis in the R statistical programming environment. It is hands-on in the sense that the book provides guidance primarily in the form of short network analysis code snippets applied to realistic network data. The results of the analyses follow immediately. All the code and data are available to the reader, so that it is easy to replicate what is shown in the book, experiment with your own data or code extensions, and thus facilitate learning.

The practical goal of the book is to demonstrate network analytic techniques in R that will be useful for a wide variety of data analysis and research goals. This includes data management, network visualization, computation of relevant network descriptive statistics, and performing mathematical, statistical, and dynamic modeling of networks. The intended audiences include students, analysts and researchers across a wide variety of disciplines, particularly the social, health, business, and engineering domains.

It is also useful to state what this book is not designed to do. First, it does not provide an in-depth treatment of network science theories or history. There are many good books, papers, training courses, and online resources available that cover this material. For good general overviews, the classic text by Wasserman and Faust (1994) is still relevant, and John Scott provides a good, more current treatment (2012). For more in-depth treatment of network science and statistical theory, see Newman (2010) or Kolaczyk (2009). Finally, two edited volumes that have good coverage of the recent history of network science as well as well-executed examples of empirical network research are Newman et al. (2006) and Scott and Carrington (2011).

Second, this book is not in any way an adequate introduction to R programming and statistical analysis. Although every attempt is made to make each code example clear and succinct, a novice R user will find some of the techniques and code syntax hard to follow. In particular, understanding R's capabilities for data management, graphics, and the object-oriented approach to statistical modeling will be very helpful for getting the most out of this user-guide.

Thus, the book is designed for the interested student, analyst, or researcher who is familiar with R and has some understanding of network science theories and methods. It could serve as a secondary text for a graduate level class in network analysis. It also could be useful as a primer for an experienced R analyst who wants to incorporate network analysis into her programming and analytic toolbox.

1.4.2 Book Roadmap

The book is organized into four main sections, which correspond to the four fundamental tasks that network analysts will spend most of their time on: data management, network visualization, network description, and network modeling. The first section has two chapters that cover both a simple introduction to basic network techniques, then a more in-depth presentation of data management issues in network analysis. The three chapters in the Visualization section cover basic network graphics layout, network graphic design suggestions, and some discussion of advanced graphics topics and techniques. The Description and Analysis section has three chapters that cover the most widely used techniques for describing important network characteristics, including actor prominence, network subgroups and communities, and handling affiliation networks. The final section, Modeling, includes four chapters that present advanced techniques for mathematical modeling, statistical modeling, modeling of dynamic networks, and network simulations. Table 1.1 presents this roadmap.

Chapter	Packages	Datasets
Introduction		FIFA_Nether, Krebs
5 number summary	statnet, sna	Moreno
Network data	statnet, network, igraph	DHHS, ICTS
Basic visualization	statnet, sna	Moreno, Bali
Graphic design	statnet, sna, igraph	Bali
Advanced graphics	arcdiagram, circlize, visNetwork, networkD3	Simpsons, Bali
Prominence	statnet, sna	DHHS, Bali
Subgroups	igraph	DHHS, Moreno, Bali
Affiliation networks	igraph	hwd
Mathematical models	igraph	lhds
Stochastic models	ergm	TCnetworks
Dynamic models	RSiena	Coevolve
Simulations	igraph	

Table 1.1 User's Guide roadmap

1.4.3 Resources

The most important resource for this user guide is a collection of network datasets that have been curated and made available to the readers of this book. Over a dozen network datasets are included in the form of an R package called UserNetR. These datasets are used throughout the book to support the coding and analysis examples. The network data included in the UserNetR package mostly come from published network studies, while a few are created to help illustrate particular analytic options. Table 1.1 lists the names of the datasets that are featured in each chapter.

The UserNetR package is maintained on GitHub, and must be downloaded and installed to make the network data available. This can be done using the following code. (The devtools package must also be installed if it is not on your system.)

```
library(devtools)
install_github("DougLuke/UserNetR")
```

Once this is done, the package must be loaded to make the various datafiles available. This can be done with the library() function, just like for any R package. This command will not always be explicitly shown throughout the book, so make sure to load the package prior to executing any of the included R code.

```
library(UserNetR)
```

Finally, the documentation for the UserNetR package can be viewed through the R help system.

```
help(package='UserNetR')
```

Part I
Network Analysis Fundamentals

Chapter 2
The Network Analysis 'Five-Number Summary'

There is nothing like looking, if you want to find something. You certainly usually find something, if you look, but it is not always quite the something you were after. (J.R.R. Tolkien – The Hobbit)

2.1 Network Analysis in R: Where to Start

How should you start when you want to do a network analysis in R? The answer to this question rests of course on the analytic questions you hope to answer, the state of the network data that you have available, and the intended audience(s) for the results of this work. The good news about performing network analysis in R is that, as will be seen in subsequent chapters, R provides a multitude of available network analysis options. However, it can be daunting to know exactly where to start.

In 1977, John Tukey introduced the five-number summary as a simple and quick way to summarize the most important characteristics of a univariate distribution. Networks are more complicated than single variables, but it is also possible to explore a set of important characteristics of a social network using a small number of procedures in R.

In this chapter, we will focus on two initial steps that are almost always useful for beginning a network analysis: simple visualization, and basic description using a 'five-number summary.' This chapter also serves as a gentle introduction to basic network analysis in R, and demonstrates how quickly this can be done.

2.2 Preparation

Similar to most types of statistical analysis using R, the first steps are to load appropriate packages (installing them first if necessary), and then making data available for the analyses. The `statnet` suite of network analysis packages will be used here for the analyses. The data used in this chapter (and throughout the rest of the book) are from the `UserNetR` package that accompanies the book. The specific dataset used here is called `Moreno`, and contains a friendship network of fourth grade students first collected by Jacob Moreno in the 1930s.

© Springer International Publishing Switzerland 2015
D.A. Luke, *A User's Guide to Network Analysis in R*, Use R!,
DOI 10.1007/978-3-319-23883-8_2

```
library(statnet)
library(UserNetR)
data(Moreno)
```

2.3 Simple Visualization

The first step in network analysis is often to just take a look at the network. Network
visualization is critical, but as Chaps. 4, 5 and 6 indicate, effective network graphics take careful planning and execution to produce. That being said, an informative
network plot can be produced with one simple function call. The only added complexity here is that we are using information about the network members' gender to
color code the nodes. The syntax details underlying this example will be covered in
greater depth in Chaps. 3, 4 and 5.

```
gender <- Moreno %v% "gender"
plot(Moreno, vertex.col = gender + 2, vertex.cex = 1.2)
```

The resulting plot makes it immediately clear how the friendship network is made
up of two fairly distinct subgroups, based on gender. A quickly produced network
graphic like this can often reveal the most important structural patterns contained in
the social network.

2.4 Basic Description

Tukey's original five-number summary was intended to describe the most important distributional characteristics of a variable, including its central tendency and
variability, using easy to produce statistical summaries. Similarly, using only a few
functions and lines of R code, we can produce a network five-number summary that
tells us how *large* the network is, how *densely* connected it is, whether the network
is made up of one or more distinct *groups*, how *compact* it is, and how *clustered* are
the network members.

2.4.1 Size

The most basic characteristic of a network is its size. The size is simply the number
of members, usually called nodes, vertices or actors. The `network.size()` function is the easiest way to get this. The basic summary of a `statnet` network object
also provides this information, among other things. The Moreno network has 33

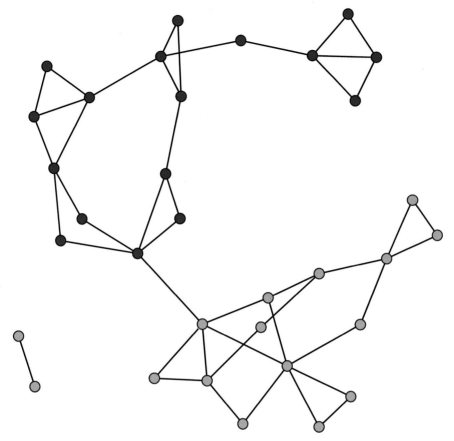

Fig. 2.1 Moreno sociogram

members, based on the network.size and summary calls. (Setting the `print.adj` to false suppresses some detailed adjacency information that can take up a lot of room.)

```
network.size(Moreno)

  ## [1] 33
```

```
summary(Moreno,print.adj=FALSE)

  ## Network attributes:
  ##    vertices = 33
  ##    directed = FALSE
  ##    hyper = FALSE
  ##    loops = FALSE
  ##    multiple = FALSE
```

```
##    bipartite = FALSE
##  total edges = 46
##    missing edges = 0
##    non-missing edges = 46
##  density = 0.0871
##
## Vertex attributes:
##
##  gender:
##    numeric valued attribute
##    attribute summary:
##    Min. 1st Qu.  Median    Mean 3rd Qu.    Max.
##    1.00    1.00    2.00    1.52    2.00    2.00
##    vertex.names:
##    character valued attribute
##    33 valid vertex names
##
## No edge attributes
```

2.4.2 Density

Of all the basic characteristics of a social network, density is among the most important as well as being one of the easiest to understand. Density is the proportion of observed ties (also called edges, arcs, or relations) in a network to the maximum number of possible ties. Thus, density is a ratio that can range from 0 to 1. The closer to 1 the density is, the more interconnected is the network.

Density is relatively easy to calculate, although the underlying equation differs based on whether the network ties are directed or undirected. An undirected tie is one with no direction. Collaboration would be a good example of an undirected tie; if A collaborates with B, then by necessity B is also collaborating with A. Directed ties, on the other hand, have direction. Money flow is a good example of a directed tie. Just because A gives money to B, does not necessarily mean that B reciprocates. For a directed network, the maximum number of possible ties among k actors is $k*(k-1)$, so the formula for density is:

$$\frac{L}{k \times (k-1)},$$

where L is the number of observed ties in the network. Density, as defined here, does not allow for ties between a particular node and itself (called a loop).

For an undirected network the maximum number of ties is $k * (k-1)/2$ because non-directed ties should only be counted once for every dyad (i.e., pair of nodes). So, density for an undirected network becomes:

$$\frac{2L}{k \times (k-1)}.$$

The information obtained in the previous section told us that the Moreno network has 33 nodes and 46 non-directed edges. We could then use R to calculate that by hand, but it is easier to simply use the `gden()` function.

```
den_hand <- 2*46/(33*32)
den_hand
```

```
## [1]  0.0871
```

```
gden(Moreno)
```

```
## [1]  0.0871
```

2.4.3 Components

A social network is sometimes split into various subgroups. Chapter 8 will describe how to use R to identify a wide variety of network groups and communities. However, a very basic type of subgroup in a network is a component. An informal definition of a component is a subgroup in which all actors are connected, directly or indirectly. The number of components in a network can be obtained with the `components` function. (Note that the meaning of components is more complicated for directed networks. See `help(components)` for more information.)

```
components(Moreno)
```

```
## [1]  2
```

2.4.4 Diameter

Although the overall size of a network may be interesting, a more useful characteristic of the network is how compact it is, given its size and degree of interconnectedness. The diameter of a network is a useful measure of this compactness. A path is the series of steps required to go from node A to node B in a network. The shortest path is the shortest number of steps required. The diameter then for an entire network is the longest of the shortest paths across all pairs of nodes. This is a measure of compactness or network efficiency in that the diameter reflects the 'worst case scenario' for sending information (or any other resource) across a network. Although social networks can be very large, they can still have small diameters because of their density and clustering (see below).

The only complicating factor for examining the diameter of a network is that it is undefined for networks that contain more than one component. A typical approach when there are multiple components is to examine the diameter of the largest component in the network. For the Moreno network there are two components (see Fig. 2.1). The smaller component only has two nodes. Therefore, we will use the larger component that contains the other 31 connected students.

In the following code the largest component is extracted into a new matrix. The geodesics (shortest paths) are then calculated for each pair of nodes using the geodist() function. The maximum geodesic is then extracted, which is the diameter for this component. A diameter of 11 suggests that this network is not very compact. It takes 11 steps to connect the two nodes that are situated the furthest apart in this friendship network.

```
lgc <- component.largest(Moreno,result="graph")
gd <- geodist(lgc)
max(gd$gdist)

  ## [1] 11
```

2.5 Clustering Coefficient

One of the fundamental characteristics of social networks (compared to random networks) is the presence of clustering, or the tendency to formed closed triangles. The process of closure occurs in a social network when two people who share a common friend also become friends themselves. This can be measured in a social network by examining its transitivity. Transitivity is defined as the proportion of closed triangles (triads where all three ties are observed) to the total number of open and closed triangles (triads where either two or all three ties are observed). Thus, like density, transitivity is a ratio that can range from 0 to 1. Transitivity of a network can be calculated using the gtrans() function. The transitivity for the 4th graders is 0.29, suggesting a moderate level of clustering in the classroom network.

```
gtrans(Moreno,mode="graph")

  ## [1] 0.286
```

In the rest of this book, we will examine in more detail how the power of R can be harnessed to explore and study the characteristics of social networks. The preceding examples show that basic plots and statistics can be easily obtained. The meaning of these statistics will always rest on the theories and hypotheses that the analyst brings to the task, as well as history and experience doing network analysis with other similar types of social networks.

Chapter 3
Network Data Management in R

Knowledge is of two kinds. We know a subject ourselves, or we know where we can find information upon it. (Samuel Johnson)

3.1 Network Data Concepts

A major advantage of using R for network analysis is the power and flexibility of the tools for accessing and manipulating the actual network data. One of the things that I often tell my quantitative methods students is that they will typically spend the majority of their time dealing with data management tasks and challenges. In fact, the time spent analyzing and modeling data is dwarfed by the time spent getting data ready for analyses. This is no different for network analysis. In fact, given the specialized nature of network data, the data management tasks loom even larger. In this chapter we cover three main topics. First, the general nature of network data is explored and defined. Second, we learn how network data objects can be created and managed in R. Finally, a number of typical network data management tasks are illustrated through a set of examples.

3.1.1 Network Data Structures

For many types of data analysis the data are stored in rectangular data structures, where rows are used to depict cases or observations, and columns depict individual variables. Spreadsheets use this type of data organization, as well as most statistics packages such as SPSS. In R one of the fundamental data types is a 'data frame,' which uses this same rectangular format.

Networks, because of their need to depict more complicated relational structures, require a different type of data storage. That is, in rectangular data structures the fundamental piece of information is an attribute (column) of a case (row). In network analysis, the fundamental piece of information is a relationship (tie) between two members of a network.

Consider the following simple example of a directed network. The network graphic itself depicts all of the information about the network. It is made up of

© Springer International Publishing Switzerland 2015
D.A. Luke, *A User's Guide to Network Analysis in R*, Use R!,
DOI 10.1007/978 3 319 23883 8_3

five nodes (named *A* through *E*), and there are a total of six directed ties. Because these are directed ties, we can call them arcs (as compared to non-directed edges). Although the network diagram is an efficient way to communicate the network information to humans, computers need to use other methods to store, access, and operate on the underlying network data.

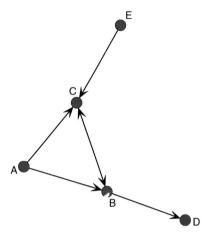

Fig. 3.1 Simple directed network

3.1.1.1 Sociomatrices

Another way to depict the network data that is more useful for computer storage is to arrange the information in a matrix. This type of matrix containing network information is a *sociomatrix*. Table 3.1 contains the sociomatrix that corresponds to Fig. 3.1. A sociomatrix is a square matrix where a 1 indicates a tie between two nodes, and a 0 indicates no tie. So in Table 3.1 we see that there is a 1 in cell 1,2–this indicates a tie going from node A to node B. The convention is that rows indicate the starting node, and columns indicate the receiving node. A sociomatrix is also sometimes called an adjacency matrix, because the 1s in the cells indicate which nodes are adjacent to one another in the network.

   If the network is non-directed (only edges instead of arcs), then the sociomatrix would be symmetric around the diagonal. Here, however, cell 2,1 has a zero, indicating that there is not an arc that goes from node B back to node A. For simple networks, there are no self-loops, where a tie connects back to its own node. So, diagonals are all zeros for simple networks.

	A	B	C	D	E
A	0	1	1	0	0
B	0	0	1	1	0
C	0	1	0	0	0
D	0	0	0	0	0
E	0	0	1	0	0

Table 3.1 Sociomatrix of the example directed network

3.1.1.2 Edge-Lists

Sociomatrices are elegant ways to depict networks, and they are a common way that many network analysis programs store and manipulate network data. In particular, many basic network algorithms are based on mathematical or statistical operations on sociomatrices. For example, to find geodesic distances between all pairs of nodes in a network the underlying sociomatrices are multiplied together (Wasserman and Faust 1994).

However, sociomatrices have one large disadvantage. As networks get larger, sociomatrices become very sparse. That is, most of the matrix will be made up of empty cells (cells with 0s). Table 3.2 shows the dramatic increase in both the size and sparseness of a sociomatrix as the network size increases, keeping the average degree constant at 3. This poses challenges for data storage, data manipulation, and data display.

Nodes	Avg. degree	Edges	Density	Empty cells
10	3	15	0.33	70
100	3	150	0.03	9,700
1,000	3	1,500	0.00	997,000

Table 3.2 Demonstration of sparse sociomatrices

Fortunately, there is another way to depict network information that avoids this problem of sociomatrices. Table 3.3 presents the *edge list* format for the example network. As its name suggests, the edge list format depicts network information by simply listing every tie in the network. Each row corresponds to a single tie, that goes *from* the node listed in the first column *to* the node listed in the second column. Although the size of the sociomatrix and the edge list matrix are similar for this small example (25 cells for the sociomatrix and 12 cells for the edge list matrix), edge lists become much more efficient for large networks. Referring back to Table 3.2, for a network with 1,000 nodes, the sociomatrix would have 1,000,000 cells. The edge list for this network, with nodes having average degree of 3, would only have 3,000 cells (1,500 edges between pairs of nodes).

From	To
A	B
A	C
B	C
B	D
C	B
E	C

Table 3.3 Edge list format for example directed network

3.1.2 Information Stored in Network Objects

Although basic matrices can be used to store some network information, R and other statistics packages use more complex data structures to contain a wide variety of network node, tie, metadata, and miscellaneous characteristics. In general, a network data object can contain up to five types of information, as listed in Table 3.4.

Type	Description	Required?
Nodes	List of nodes in network, along with node labels	Required
Ties	List of ties in the network	Required
Node attributes	Attributes of the nodes	Optional
Tie attributes	Attributes of the ties	Optional
Metadata	Other information about the entire network	Depends

Table 3.4 Types of information contained in network data objects

First, a network data object must know which objects belong to the network, these are generally known as nodes (in `statnet` they are called vertices). The second required component in a network object is the list of ties that connect the nodes to one another. Without these two types of information, the data object is not really a network object. In addition to node and tie listings, network data objects will often be able to store characteristics of those nodes and ties. For example, if the nodes in the network are people, then basic information on those peoples such as gender or income could be contained in the data object. Similarly, ties themselves may have characteristics such as strength or valence (e.g., positive vs. negative). Finally, network data objects may also contain metadata about the whole network or other information that may be relevant or useful when accessing or analyzing the data. For example, `statnet` stores global information about the network as metadata, including whether the network is directed, whether loops are allowed, and whether the network is bipartite.

3.2 Creating and Managing Network Objects in R

Given R's object-oriented design, it is not surprising that the main way that R expects to access network data is through some type of a network data object. As part of the statnet suite of packages, the network package defines a network class that is an object structure designed to hold network data. Although statnet can recognize relational data that are stored in basic matrices or data frames, much of the power and flexibility of R's network analyses is unlocked when using network data objects. For more detailed information about network objects in statnet, see Butts (2008).

3.2.1 Creating a Network Object in statnet

To create a network object, the identically-named network() function is called. This function has a number of options, but the most common way to use it is to feed relational data to it–typically an adjacency matrix or edge list. To see how this works we will continue with the example directed network from Fig. 3.1. First, we will create a network using an adjacency matrix.

```
netmat1 <- rbind(c(0,1,1,0,0),
                 c(0,0,1,1,0),
                 c(0,1,0,0,0),
                 c(0,0,0,0,0),
                 c(0,0,1,0,0))
rownames(netmat1) <- c("A","B","C","D","E")
colnames(netmat1) <- c("A","B","C","D","E")
net1 <- network(netmat1,matrix.type="adjacency")
class(net1)

  ## [1] "network"

summary(net1)

  ## Network attributes:
  ##    vertices = 5
  ##    directed = TRUE
  ##    hyper = FALSE
  ##    loops = FALSE
  ##    multiple = FALSE
  ##    bipartite = FALSE
  ##  total edges = 6
  ##    missing edges = 0
  ##    non-missing edges = 6
  ##  density = 0.3
```

```
##
## Vertex attributes:
##   vertex.names:
##     character valued attribute
##     5 valid vertex names
##
## No edge attributes
##
## Network adjacency matrix:
##   A B C D E
## A 0 1 1 0 0
## B 0 0 1 1 0
## C 0 1 0 0 0
## D 0 0 0 0 0
## E 0 0 1 0 0
```

The results of the class() and summary() calls show that we have successfully created a new network object. Also, this demonstrates that if the matrix has identical row and column names, they will be used as the labels for the nodes. We can also see that this is the same network as the earlier example by plotting it (Fig. 3.2).

```
gplot(net1, vertex.col = 2, displaylabels = TRUE)
```

The same network can be created using an edge list format. This will often be more convenient than adjacency matrices. Not only are edge lists smaller than sociomatrices, but network data are often obtained naturally in this format. For example, email communications can be analyzed as networks, where each email corresponds to a tie from the email sender to the receiver. This leads easily to edge list node pairs.

```
netmat2 <- rbind(c(1,2),
                 c(1,3),
                 c(2,3),
                 c(2,4),
                 c(3,2),
                 c(5,3))
net2 <- network(netmat2,matrix.type="edgelist")
network.vertex.names(net2) <- c("A","B","C","D","E")
summary(net2)

## Network attributes:
##   vertices = 5
##   directed = TRUE
##   hyper = FALSE
##   loops = FALSE
```

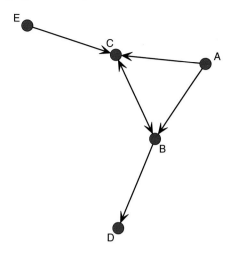

Fig. 3.2 Plot of new network object

```
##    multiple = FALSE
##    bipartite = FALSE
##  total edges = 6
##    missing edges = 0
##    non-missing edges = 6
##  density = 0.3
##
## Vertex attributes:
##   vertex.names:
##     character valued attribute
##     5 valid vertex names
##
## No edge attributes
##
## Network adjacency matrix:
##   A B C D E
## A 0 1 1 0 0
## B 0 0 1 1 0
## C 0 1 0 0 0
## D 0 0 0 0 0
## E 0 0 1 0 0
```

This produces the same network as before. Notice that the edgelist was provided in the form of node ID numbers. To label the nodes properly, we used a special vertex attribute constructor, network.vertex.names.

We have seen that to create network objects in R we can use a workflow that takes data in a number of basic matrix formats and transforms them into the network class object. However, `statnet` also includes a number of tools that allow you to reverse this workflow, by coercing network data into other matrix formats.

```
as.sociomatrix(net1)

   ##    A B C D E
   ## A 0 1 1 0 0
   ## B 0 0 1 1 0
   ## C 0 1 0 0 0
   ## D 0 0 0 0 0
   ## E 0 0 1 0 0

class(as.sociomatrix(net1))

   ## [1] "matrix"
```

A more general coercion function is `as.matrix()`. It can be used to produce a sociomatrix or an edgelist matrix.

```
all(as.matrix(net1) == as.sociomatrix(net1))

   ## [1] TRUE

as.matrix(net1,matrix.type = "edgelist")

   ##          [,1] [,2]
   ## [1,]        1    2
   ## [2,]        3    2
   ## [3,]        1    3
   ## [4,]        2    3
   ## [5,]        5    3
   ## [6,]        2    4
   ## attr(,"n")
   ## [1] 5
   ## attr(,"vnames")
   ## [1] "A" "B" "C" "D" "E"
```

This ability to go back and forth between network objects and more fundamental data structures such as sociomatrices and edgelist matrices gives the analyst great power and flexibility when managing network data. We will take advantage of these tools later in this chapter as well as throughout the book.

3.2.2 Managing Node and Tie Attributes

One of the major advantages of using network objects when doing network analysis in R rather than using simpler matrix objects is the ability to store additional attribute information about the nodes and ties within the same network object. The analyst typically knows much more about the members of a network than just the simple list of nodes and ties. These node or tie characteristics can be used in network visualization (see Chap. 5), network description, and network modeling (Chap. 11).

For both nodes and ties, statnet provides a set of functions that can be used to create, delete, access, and list any attribute information of relevance. These functions have a lot of capabilities, see help(attribute.methods) for more details.

3.2.2.1 Node Attributes

In the following example we use two different methods to set a pair of node attributes (called vertex attributes by statnet). The first example uses the more formal method to assign gender codes to the nodes in net1. The second example uses a shorthand method to assign a numeric vector as an attribute. In this case we are storing the sum of the indegrees and outdegrees of each node as a new vertex attribute.

```
set.vertex.attribute(net1, "gender", c("F", "F", "M",
    "F", "M"))
net1 %v% "alldeg" <- degree(net1)
list.vertex.attributes(net1)

    ## [1] "alldeg"         "gender"         "na"
    ## [4] "vertex.names"

summary(net1)

    ## Network attributes:
    ##    vertices = 5
    ##    directed = TRUE
    ##    hyper = FALSE
    ##    loops = FALSE
    ##    multiple = FALSE
    ##    bipartite = FALSE
    ##  total edges = 6
    ##    missing edges = 0
    ##    non-missing edges = 6
    ##  density = 0.3
    ##
    ## Vertex attributes:
    ##
    ##  alldeg:
    ##    numeric valued attribute
    ##    attribute summary:
    ##    Min. 1st Qu.  Median   Mean 3rd Qu.   Max.
```

```
##      1.0      1.0      2.0      2.4      4.0      4.0
##
##  gender:
##    character valued attribute
##    attribute summary:
## F M
## 3 2
##    vertex.names:
##    character valued attribute
##    5 valid vertex names
##
## No edge attributes
##
## Network adjacency matrix:
##    A B C D E
## A 0 1 1 0 0
## B 0 0 1 1 0
## C 0 1 0 0 0
## D 0 0 0 0 0
## E 0 0 1 0 0
```

In this example, we see that information obtained outside of the network (i.e., gender) or information obtained from the network itself (i.e., degree) can be used as node attributes. Once node attributes have been set, they can be examined with the `list.vertex.attributes` command (note the plural). Also, the summary of the network will provide some basic information about any stored attributes.

To see the actual values stored in a vertex attribute, you can use the following two equivalent methods.

```
get.vertex.attribute(net1, "gender")
```

```
## [1] "F" "F" "M" "F" "M"
```

```
net1 %v% "alldeg"
```

```
## [1] 2 4 4 1 1
```

3.2.2.2 Tie Attributes

Information about tie characteristics can also be stored and managed in the network objects, using the similarly named `set.edge.attributes` and `get.edge.attributes` functions. In the following example we create a new edge attribute that contains a random number for each edge in the network, and then access that information.

```
list.edge.attributes(net1)

  ## [1] "na"

set.edge.attribute(net1,"rndval",
                   runif(network.size(net1),0,1))
list.edge.attributes(net1)

  ## [1] "na"       "rndval"

summary(net1 %e% "rndval")

  ##     Min. 1st Qu.  Median   Mean 3rd Qu.    Max.
  ##    0.163   0.165   0.220  0.382   0.476   0.980

summary(get.edge.attribute(net1,"rndval"))

  ##     Min. 1st Qu.  Median   Mean 3rd Qu.    Max.
  ##    0.163   0.165   0.220  0.382   0.476   0.980
```

A more typical situation where you will want to create a new edge attribute is when you are creating or working with valued networks. A valued network is one where the network tie has some numeric value. For example, a resource exchange network may include not just whether this is a flow of money from one node to another, but the actual amount of that money. In statnet, the actual values of the valued ties are stored in an edge attribute. To see how this works, consider our example network now as a friendship network, where the five network members were asked to indicate how much they liked one another, on a scale of 0 (not at all) to 3 (very much). The following example shows how we would proceed from the raw valued sociomatrix to storing the values in an edge attribute called 'like.'

```
netval1 <- rbind(c(0,2,3,0,0),
                 c(0,0,3,1,0),
                 c(0,1,0,0,0),
                 c(0,0,0,0,0),
                 c(0,0,2,0,0))
netval1 <- network(netval1,matrix.type="adjacency",
                   ignore.eval=FALSE,names.eval="like")
network.vertex.names(netval1) <- c("A","B","C","D","E")
list.edge.attributes(netval1)

  ## [1] "like" "na"

get.edge.attribute(netval1, "like")

  ## [1] 2 1 3 3 2 1
```

The key here are the ignore.eval and names.eval options. These two options, as set here, tell the network function to evaluate the actual values in the sociomatrix, and store those values in a new edge attribute called 'like.' Once values are stored in an edge attribute, the original valued matrix can be restored using as option of the `as.sociomatrix` coercion function.

```
as.sociomatrix(netval1)

  ##   A B C D E
  ## A 0 1 1 0 0
  ## B 0 0 1 1 0
  ## C 0 1 0 0 0
  ## D 0 0 0 0 0
  ## E 0 0 1 0 0

as.sociomatrix(netval1,"like")

  ##   A B C D E
  ## A 0 2 3 0 0
  ## B 0 0 3 1 0
  ## C 0 1 0 0 0
  ## D 0 0 0 0 0
  ## E 0 0 2 0 0
```

3.2.3 Creating a Network Object in `igraph`

The other major R package that can be used to store and manipulate network data is `igraph`, which is a comprehensive set of network data management and analytic tools that have been implemented in R, Python, and C/C++. More information can be obtained at `igraph.org`.

To start working with `igraph`, the package needs to be installed and loaded. It contains a number of functions that have the same names as those found in the `statnet` suite of packages, so it is a good idea to detach `statnet` before loading `igraph`.

```
detach(package:statnet)
library(igraph)
```

For the most part, `igraph` can be used to store and access network, node, and edge information in similar ways as the `network` package. In particular, `igraph` network objects (called 'graphs') can be created from more basic sociomatrix or edge list data structures.

```
inet1 <- graph.adjacency(netmat1)
class(inet1)

  ## [1] "igraph"
```

```
summary(inet1)

  ## IGRAPH DN-- 5 6 --
  ## + attr: name (v/c)

str(inet1)

  ## IGRAPH DN-- 5 6 --
  ## + attr: name (v/c)
  ## + edges (vertex names):
  ## [1] A->B A->C B->C B->D C->B E->C
```

The summary information from an `igraph` graph object is slightly more cryptic than from a `statnet` network. After the 'IGRAPH' tag is listed (indicating that this is an `igraph` object), a series of codes are presented. In this case the 'D' indicates a directed graph, and the 'N' indicates that the vertices are named. Other codes might appear that would designate whether the graph is weighted (i.e., valued) or bipartite. After these codes the number of vertices (5) and edges (6) are then displayed. See the help entry for `summary.igraph` for more details. The `str()` function provides slightly more information, including the edge list.

Similarly, an `igraph` graph object can be created from an edge list.

```
inet2 <- graph.edgelist(netmat2)
summary(inet2)

  ## IGRAPH D--- 5 6 --
```

Node and tie attributes can be created, accessed, and transformed in similar ways as within `statnet`. (In fact, management of node and tie attributes is somewhat easier in `igraph` because of the underlying elegance of the accessor functions.) To create and use node attributes, the `V()` vertex accessor function is used. Similarly, to manage edge attributes, the `E()` edge accessor function is used. In this example we use these functions to set names for the nodes, and to set edge values for the observed ties.

```
V(inet2)$name <- c("A","B","C","D","E")
E(inet2)$val <- c(1:6)
summary(inet2)

  ## IGRAPH DN-- 5 6 --
  ## + attr: name (v/c), val (e/n)

str(inet2)

  ## IGRAPH DN-- 5 6 --
  ## + attr: name (v/c), val (e/n)
  ## + edges (vertex names):
  ## [1] A->B A->C B->C B->D C->B E->C
```

3.2.4 Going Back and Forth Between statnet *and* igraph

There will be times when you will want to use statnet network functions on
network data stored in an igraph graph object, and vice versa. To facilitate this,
the intergraph package can be used to transform network data objects between
the two formats. In the following example, we transform the net1 data into the
igraph format using the asIgraph function. If we wanted to go in the opposite
direction, we would use asNetwork.

```
library(intergraph)
class(net1)

  ## [1] "network"

net1igraph <- asIgraph(net1)
class(net1igraph)

  ## [1] "igraph"

str(net1igraph)

  ## IGRAPH D--- 5 6 --
  ## + attr: alldeg (v/n), gender (v/c), na
  ## | (v/l), vertex.names (v/c), na (e/l),
  ## | rndval (e/n)
  ## + edges:
  ## [1] 1->2 3->2 1->3 2->3 5->3 2->4
```

3.3 Importing Network Data

Importing raw data into R for subsequent network analyses is relatively straightforward, as long as the external data are in edge list, adjacency list, or sociomatrix
form (or can easily be transformed into such). This example creates an edge list that
corresponds to the same example network from Sect. 3.2.1 and then saves it as an
external CSV file. This file is then read in using read.csv and then turned into a
network data object.

```
detach("package:igraph", unload=TRUE)
library(statnet)


netmat3 <- rbind(c("A","B"),
                 c("A","C"),
                 c("B","C"),
```

```
                          c("B","D"),
                          c("C","B"),
                          c("E","C"))
net.df <- data.frame(netmat3)
net.df

  ##    X1 X2
  ## 1  A  B
  ## 2  A  C
  ## 3  B  C
  ## 4  B  D
  ## 5  C  B
  ## 6  E  C

write.csv(net.df, file = "MyData.csv",
          row.names = FALSE)
net.edge <- read.csv(file="MyData.csv")
net_import <- network(net.edge,
                         matrix.type="edgelist")
summary(net_import)

  ## Network attributes:
  ##   vertices = 5
  ##   directed = TRUE
  ##   hyper = FALSE
  ##   loops = FALSE
  ##   multiple = FALSE
  ##   bipartite = FALSE
  ## total edges = 6
  ##   missing edges = 0
  ##   non-missing edges = 6
  ## density = 0.3
  ##
  ## Vertex attributes:
  ##   vertex.names:
  ##     character valued attribute
  ##     5 valid vertex names
  ##
  ## No edge attributes
  ##
  ## Network adjacency matrix:
  ##    A B C D E
  ## A  0 1 1 0 0
  ## B  0 0 1 1 0
  ## C  0 1 0 0 0
  ## D  0 0 0 0 0
```

```
## E 0 0 1 0 0
```

```
gden(net_import)
```

```
## [1] 0.3
```

The `network` package in the `statnet` suite can read in external network data that are in Pajek format (either Pajek .net or .paj files), using the `read.paj()` function. The `igraph` package can also import Pajek files, as well as a few other formats including GraphML and UCINet DL files.

3.4 Common Network Data Tasks

The preceding sections covered the basic information needed to create and manage network data objects in R. However, the data managements tasks for network analysis do not end there. There are any number of network analytic challenges that will require more sophisticated data management and transformation techniques. In the rest of this chapter, two such examples are covered: preparing subsets of network data for analysis by filtering on node and edge characteristics, and turning directed networks into non-directed networks.

3.4.1 Filtering Networks Based on Vertex or Edge Attribute Values

It is quite common to want to examine a subset of a network, either for quick visualization or for further analyses. There are many ways to define or identify interesting subnetworks in a larger network, and Chap. 8 covers many of them. However, as a basic data management task, you can filter a network based on values contained either in edge attributes or vertex attributes. For both of these cases, you will delete either the nodes or the edges, based on selection criteria that you set.

3.4.1.1 Filtering Based on Node Values

If a network object contains node characteristics, stored as vertex attributes, this information can be used to select a new subnetwork for analysis. In our example network we have the `gender` vertex attribute, so if you wanted to look at the subnetwork made up of females, you would use the following code (after switching back from `igraph` to `statnet`).

```
n1F <- get.inducedSubgraph(net1,
           which(net1 %v% "gender" == "F"))
n1F[,]

##    A B D
## A 0 1 0
## B 0 0 1
## D 0 0 0
```

The get.inducedSubgraph() function returns a new network object that is filtered based on the vertex attribute criteria. This works because the %v% operator returns a list of vertex ids.

```
gplot(n1F,displaylabels=TRUE)
```

The same process can work with numeric node characteristics. The following code will plot the subset of the example network who all have degree greater than or equal to 2. (But note that the nodes in the new subnetwork will of course not have the same original degree values!) This works the same way but uses the %s% operator, which is a shortcut for the get.inducedSubgraph function (Fig. 3.3).

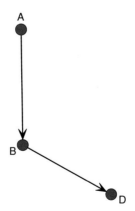

Fig. 3.3 Female subnetwork

```
deg <- net1 %v% "alldeg"
n2 <- net1 %s% which(deg > 1)
```

```
gplot(n2,displaylabels=TRUE)
```

3.4.1.2 Removing Isolates

Another common filtering task with networks is to examine the network after removing all the isolates (i.e., nodes with degree of 0). We could use the `get.inducedSubgraph` function from the previous section, but given that we want to delete certain nodes we can take a more direct approach (Fig. 3.4).

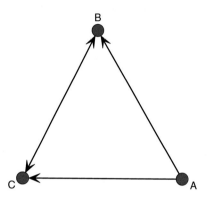

Fig. 3.4 High degree subnetwork

For this short example, we will use the ICTS network dataset, which is available as part of the `UserNetR` package that accompanies this book. The members of this network are scientists, and they have a tie if they worked together on a scientific grant submission. Using the `isolates()` function, we can see that this network has a fair number of isolated nodes.

```
data(ICTS_G10)
gden(ICTS_G10)

  ## [1] 0.0112

length(isolates(ICTS_G10))

  ## [1] 96
```

The `isolates()` function returns a vector of vertex IDs. This can be fed to the `delete.vertices()` function. However, unlike most R functions we have seen, `delete.vertices()` does not return an object, but it directly operates on the network that is passed to it. For that reason, it is safer to work on a copy of the object.

```
n3 <- ICTS_G10
delete.vertices(n3,isolates(n3))
gden(n3)
```

```
  ## [1] 0.0173
```

```
length(isolates(n3))
```

```
  ## [1] 0
```

3.4.1.3 Filtering Based on Edge Values

A social network often contains valued ties. For example, a resource exchange network may list not only who exchanges money (or some other resource) with each
other, but the amount of money. Remember that in statnet information about ties
is stored in edge attributes (see Sect. 3.2.2). When a network has valued ties, it is not
unusual to want to examine the part of the network that only has certain values for
those ties. For example, you might want to visualize or analyze only those persons
in the resource exchange network who have given or received over a certain amount
of money. For this you will need to filter the network using tie values contained in
the appropriate edge attribute.

For this next example we will use a larger, more realistic social network. The
DHHS Collaboration Network (DHHS) contains network data from a study of the
relationships among 54 tobacco control experts working in 11 different agencies
in the Department of Health and Human Services in 2005. The main relationship
included in this dataset is collaboration – two members have a tie if they worked
together in the past year. This tie is valued to capture differences in the strength of
the collaboration. Specifically, the collaboration tie could take on one of four values:
(1) Shared information only; (2) Worked together informally; (3) Worked together
formally on a project; and (4) Worked together formally on multiple projects.

We can see that the raw network is relatively dense, and because of that the
network structure is somewhat hard to interpret when plotted (Fig. 3.5).

```
data(DHHS)
d <- DHHS
gden(d)
```

```
  ## [1] 0.312
```

```
op <- par(mar = rep(0, 4))
gplot(d,gmode="graph",edge.lwd=d %e% 'collab',
      edge.col="grey50",vertex.col="lightblue",
      vertex.cex=1.0,vertex.sides=20)
par(op)
```

The graphic hard to interpret partly because of the high density, as well as having
some edge widths being thicker based on the value of the 'collab' attribute. We may
have a more interesting network graph (and one that is easier to interpret), if we
only examine the network ties for formal collaboration. That is, we can filter the

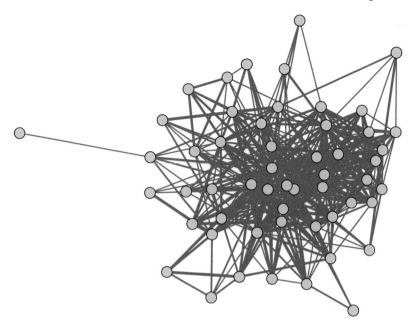

Fig. 3.5 DHHS collaborations

original network and show only those ties where collaboration is coded 3 or higher.
To understand how edge filtering works, it is important to remember how valued ties
are stored in a network object. The ties themselves are stored as a binary indicator
in the network object, while the values of those ties are stored in an edge attribute.
We can see how this works for the DHHS Collaboration network. First, we examine
the network ties for the first six members of the network. Then we determine where
the collaboration values are stored, and then use that to view the tie values for the
same set of six actors.

```
as.sociomatrix(d)[1:6,1:6]

  ##            ACF-1 ACF-2 AHRQ-1 AHRQ-2 AHRQ-3 AHRQ-4
  ## ACF-1          0     1      0      0      0      0
  ## ACF-2          1     0      0      0      0      0
  ## AHRQ-1         0     0      0      1      1      1
  ## AHRQ-2         0     0      1      0      1      1
  ## AHRQ-3         0     0      1      1      0      1
  ## AHRQ-4         0     0      1      1      1      0

list.edge.attributes(d)

  ## [1] "collab" "na"

as.sociomatrix(d,attrname="collab")[1:6,1:6]
```

```
##             ACF-1 ACF-2 AHRQ-1 AHRQ-2 AHRQ-3 AHRQ-4
## ACF-1          0     1      0      0      0      0
## ACF-2          1     0      0      0      0      0
## AHRQ-1         0     0      0      3      3      3
## AHRQ-2         0     0      3      0      3      2
## AHRQ-3         0     0      3      3      0      3
## AHRQ-4         0     0      3      2      3      0
```

The summary of the network object tells us that there are 447 ties in the DHHS network. We can easily see the distribution of tie values.

```
table(d %e% "collab")

##
##   1   2   3   4
## 163 111  94  79
```

This indicates that of the 447 ties, 163 are informal sharing (1), 111 are informal (2), 94 are formal on a single project (3), and the final 79 ties are between DHHS members who have worked together formally on multiple projects (4). Now we can filter the edges to only include formal collaboration ties. This takes three steps. First, a valued sociomatrix is created that contains the tie values stored in the 'collab' edge attribute. Then we filter out the ties that we want to ignore. In this case the ties that are coded 1 and 2 are replaced with 0s. Then, we create a new network based on the filtered sociomatrix. The key here is that a tie will be created anywhere a non-zero value is found in d.val. Also, by using the ignore.eval and names.eval options we store the retained edge values in an edge attribute called 'collab.'

```
d.val <- as.sociomatrix(d, attrname="collab")
d.val[d.val < 3] <- 0
d.filt <- as.network(d.val, directed=FALSE,
                     matrix.type="a", ignore.eval=FALSE,
                     names.eval="collab")
```

We can see that the new network has the same number of actors, but only 173 ties (corresponding to the original numbers for the 3 and 4-levels of collab). Also, not surprisingly, the density is now much lower.

```
summary(d.filt, print.adj=FALSE)

## Network attributes:
##   vertices = 54
##   directed = FALSE
##   hyper = FALSE
##   loops = FALSE
##   multiple = FALSE
##   bipartite = FALSE
```

```
##   total edges = 173
##     missing edges = 0
##     non-missing edges = 173
##   density = 0.121
##
## Vertex attributes:
##    vertex.names:
##      character valued attribute
##      54 valid vertex names
##
## Edge attributes:
##
##   collab:
##      numeric valued attribute
##      attribute summary:
##      Min. 1st Qu.  Median   Mean 3rd Qu.    Max.
##      3.00    3.00    3.00   3.46    4.00    4.00
```

```
gden(d.filt)
```

```
## [1] 0.121
```

Now when the network is plotted we can examine a smaller set of ties for important structural information (Fig. 3.6).

```
op <- par(mar = rep(0, 4))
gplot(d.filt,gmode="graph",displaylabels=TRUE,
      vertex.col="lightblue",vertex.cex=1.3,
      label.cex=0.4,label.pos=5,
      displayisolates=FALSE)
par(op)
```

Note that the gplot() function itself has a limited ability to display only the ties that exceed some lower threshold, using the thresh option. For example, this command will display the same network as the previous code, without having to go through the steps to create a new filtered network (results not shown here). Note that for this to work a valued sociomatrix has to be passed to gplot, not an actual network object.

```
op <- par(mar = rep(0, 4))
d.val <- as.sociomatrix(d,attrname="collab")
gplot(d.val,gmode="graph",thresh=2,
      vertex.col="lightblue",vertex.cex=1.3,
      label.cex=0.4,label.pos=5,
      displayisolates=FALSE)
par(op)
```

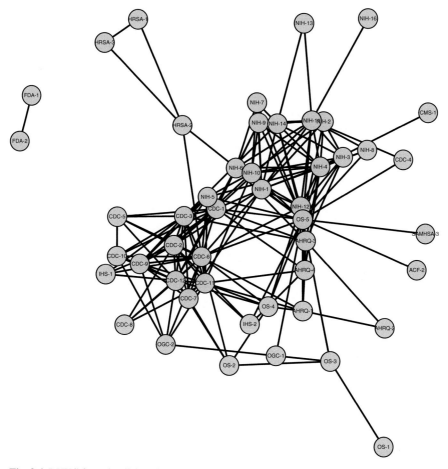

Fig. 3.6 DHHS formal collaborations

3.4.2 Transforming a Directed Network to a Non-directed Network

It is often the case that even though the raw data in a network analysis is made up of directed ties, the analyst wishes to consider the data as non-directed. This could happen for several reasons. First, although the network relationship is non-directed, the data collection procedures may result in directed ties. For example, in a survey of collaboration among organizational representatives, even though collaboration is non-directed (if agency A is collaborating with agency B, then B is also collaborating with A), the raw data matrices are not likely to be perfectly symmetric. That is, in self-report data, there may be error in the data or pairs or respondents may not agree with each other on collaboration status. Respondent A may believe that agency A collaborates with agency B, but respondent B may not believe the two agencies are collaborating. In any case, you end up with a directed network that you wish to 'fix' by transforming it into a non-directed network.

You may also wish to transform directed ties into non-directed ties for conceptual reasons, and not simply to fix data disagreements. For example, in studying trust relationships you collect data on directed perceptions of trust. Here a tie between A and B indicates that A trusts B. This is directed, in the sense that just because A trusts B, that does not mean that B trusts A in return. However, you may wish to analyze this in a non-directed sense, where an edge exists between two actors if there is any trust relationship between the pair. So whether A trusts B, B trusts A, or even if there is a reciprocal trusting relationship between A and B, then you would treat A and B as having a trust relationship where you are ignoring the directionality of the trust.

For either of these reasons, R makes it easy to transform a directed network into a non-directed network. To do this you can use the symmetrize() function. The name of the function should remind you that when network data are stored in a sociomatrix, if the data are symmetric around the diagonal that indicates that the ties are non-directed.

```
net1mat <- symmetrize(net1,rule="weak")
net1mat
```

```
##          [,1] [,2] [,3] [,4] [,5]
## [1,]      0    1    1    0    0
## [2,]      1    0    1    1    0
## [3,]      1    1    0    0    1
## [4,]      0    1    0    0    0
## [5,]      0    0    1    0    0
```

```
net1symm <- network(net1mat,matrix.type="adjacency")
network.vertex.names(net1symm) <- c("A","B","C","D","E")
summary(net1symm)
```

```
## Network attributes:
##    vertices = 5
##    directed = TRUE
##    hyper = FALSE
##    loops = FALSE
##    multiple - FALSE
##    bipartite = FALSE
##  total edges = 10
##    missing edges = 0
##    non-missing edges = 10
##  density = 0.5
##
## Vertex attributes:
##    vertex.names:
```

```
##      character valued attribute
##      5 valid vertex names
##
## No edge attributes
##
## Network adjacency matrix:
##    A B C D E
## A 0 1 1 0 0
## B 1 0 1 1 0
## C 1 1 0 0 1
## D 0 1 0 0 0
## E 0 0 1 0 0
```

The symmetrize procedure is relatively straightforward, except it returns a sociomatrix (or, optionally, an edgelist). So we need to then turn it into a network object, as we have done previously. The 'rule' option gives you four different choices on how to symmetrize the ties. The 'weak' rule corresponds to a Boolean 'OR' condition where a tie is created between nodes i and j if there is a directed tie either from i to j or from j to i. There is also a 'strong' rule, corresponding to a Boolean 'AND' where a tie is created between i and j only if there are directed ties from i to j and j to i. This creates a symmetric network where the only ties preserved are the fully reciprocated ties.

Part II
Visualization

Chapter 4
Basic Network Plotting and Layout

Above all else, show the data. (Edward R. Tufte, *The Visual Display of Quantitative Information*)

4.1 The Challenge of Network Visualization

As suggested in Chap. 2, producing and examining a network plot is often one of the first steps in network analysis. The overall purpose of a network graphic (as with any information graphic) is to highlight the important information contained in the underlying data. However, there are innumerable ways to visually layout network nodes and ties in two-dimensional space, as well as using graphical elements (e.g., node size, line color, figure legend, etc.) to communicate the story in the network data. In the next three chapters we go over basic principles of effective network graph design, and how to produce effective network visualizations in R.

An effective network graphic will convey the important information in a social network, such as the overall structure, location of important actors in the network, presence of distinctive subgroups, etc. At the same time, the graphic should do its best to minimize irrelevant information. For example, tie length in a network graphic is arbitrary in the sense that the length of a tie is not meaningful. An effective network figure will be designed and laid out in a way that minimizes the chance that a viewer will misinterpret the meaning of tie lengths.

The purpose of this chapter is to introduce basic plotting techniques for networks in R, and discuss the various options for specifying the layout of the network on the screen or page. The following example shows how interpretation of a network graphic can be impeded or enhanced by its basic layout.

```
data(Moreno)
op <- par(mar = rep(0, 4),mfrow=c(1,2))
plot(Moreno,mode="circle",vertex.cex=1.5)
plot(Moreno,mode="fruchtermanreingold",vertex.cex=1.5)
par(op)
```

At first glance it may appear that the figures are showing two quite different networks. In fact, they are two different visual representations of the same underlying

© Springer International Publishing Switzerland 2015
D.A. Luke, *A User's Guide to Network Analysis in R*, Use R!,
DOI 10.1007/978 3 319 23883 8_4

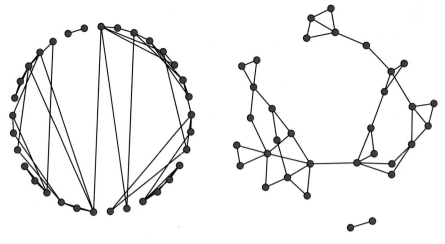

Fig. 4.1 Same network, different layouts

social network, in this case the friendship ties among a 4th grade class. Despite representing the same network data, the righthand figure is easier for us to interpret. In particular, it is much easier to see that the network is made up of two separate components, and that the large component has two fairly distinct cohesive subgroups. That is, the important structural characteristics of the network are easier to determine with the second layout compared to the first.

Although it is possible to lay out a network in 3D-space, the vast majority of network visualizations are two-dimensional. Nodes are represented by shapes, typically circles, and ties are represented by straight or sometimes curved lines. The lines themselves can be tricky to interpret for somebody new to network visualization. In particular, the length of the line has no real meaning. Consider the following two graphs, which display the same simple network (Fig. 4.2). At a quick glance it might appear that node D is further away from B and C in the second graph. But the ties simply indicate which nodes are *adjacent* to one another, so the length of each line does not communicate any substantive information.

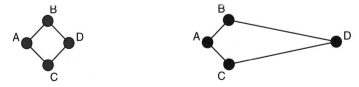

Fig. 4.2 Line length is arbitrary

However, as the Moreno 4th grade friendship network example illustrated (Fig. 4.1), despite the arbitrary nature of some of the layout elements, the way a network is depicted in a graphic can enhance or obscure other important structural information.

This is the fundamental challenge of network visualization: to reveal important structural characteristics of the network without distortion or as Edward Tufte stated, *The minimum we should hope for with any display technology is that it should do no harm* (Tufte 1990).

4.2 The Aesthetics of Network Layouts

Although there are not in fact an infinite number of ways to display a network on a screen, the number of possibilities might as well be. (For example, consider a moderate sized network of 50 nodes, and a display grid of 10 by 10. In actuality, the display grid would be much larger than this. The first node in the network could be placed in any one of 100 positions, the 2nd node in 99 positions, and so on. In this example, there are 3.1×10^{93} different possible network layouts.) Most of the possibilities will produce ugly or confusing layouts, therefore there must be some way to pick a layout that has a better than average chance of being visually acceptable.

Fortunately, network and visualization scientists have studied what makes network graph layouts easier to understand and interpret. What has emerged from this line of work is a set of aesthetic principles that can be used to more effectively display networks. Network graphics are easier to understand if they follow as much as possible the following five guidelines:

- Minimize edge crossings.
- Maximize the symmetry of the layout of nodes.
- Minimize the variability of the edge lengths.
- Maximize the angle between edges when they cross or join nodes.
- Minimize the total space used for the network display.

A large number of approaches have been developed for automatic layout of network graphics. One general class of algorithms, called *force-directed*, has proven to be a flexible and powerful approach to automatic network layouts. These algorithms work iteratively to minimize the total energy in a network, where the energy can be defined in a number of ways. A popular approach is to have connected nodes have a spring-like attractive force, while simultaneously assigning repulsive forces to all pairs of nodes. The *springs* in this algorithm act to pull connected nodes closer to one another, while the repulsive forces push unconnected nodes away from each other. The resulting network system will move around and oscillate for a while before settling into a steady state that tends to minimize the energy in the network system. This describes how the algorithm works, but the remarkable feature is that the resulting network graph tends to produce displays that are aesthetically pleasing, in the sense described above (Fruchterman and Reingold 1991).

To see the positive results of using one of these algorithms, consider the comparison in Fig. 4.3. On the left-hand side, the Moreno network is displayed randomly. On the right-hand side we are using the Fruchterman-Reingold algorithm for the network display. Fruchterman and Reingold introduced one of the first force-directed

network display algorithms, and it is still very widely used. In fact, it is the default
algorithm used by the `statnet` network plotting functions. On the right-hand side
the nodes are displayed more symmetrically, there are relatively fewer edge crossings, and the tie lengths are more uniform. All of this makes it easier to interpret the
structural information contained in the network.

```
op <- par(mar = c(0,0,4,0),mfrow=c(1,2))
gplot(Moreno,gmode="graph",mode="random",
      vertex.cex=1.5,main="Random layout")
gplot(Moreno,gmode="graph",mode="fruchtermanreingold",
      vertex.cex=1.5,main="Fruchterman-Reingold")
par(op)
```

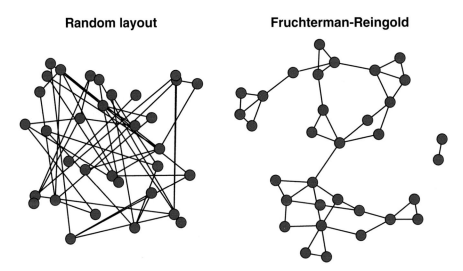

Fig. 4.3 Moreno network-random vs. Fruchterman-Reingold layouts

As stated above, a force-directed algorithm works by iteratively adjusting the
overall network layout until some measure of overall network energy is minimized.
The details of this are usually not of interest, but to see how this works in practice
consider Fig. 4.4, which displays the Bali terrorist network. Starting from a circle
layout, it shows how the Fruchterman-Reingold layout algorithm works through
successive iterations, from 0 (the starting circle) to 50.

The Fruchterman-Reingold algorithm, along with other force-directed approaches,
are iterative and non-deterministic. That means that each time you run the plotting
algorithm you will not get the exact same layout. However, you will get a layout
that tends to be symmetrical, minimize edge crossings, etc.

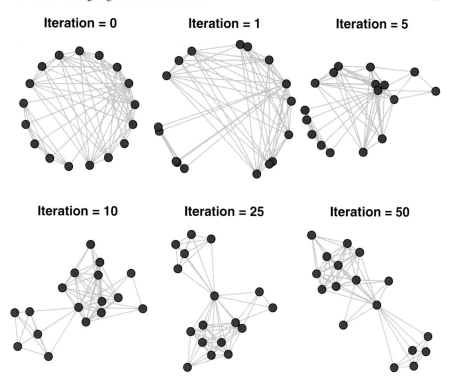

Fig. 4.4 Iterative Fruchterman-Reingold algorithm

4.3 Basic Plotting Algorithms and Methods

Network visualization in statnet is handled by two closely related functions, plot and gplot. The latter has more layout options, so it may be more generally useful. To use a different layout algorithm, it is as simple as specifying the appropriate layout option. Figure 4.5 shows six of the layout options for the gplot function.

```
op <- par(mar=c(0,0,4,0),mfrow=c(2,3))
gplot(Bali,gmode="graph",edge.col="grey75",
      vertex.cex=1.5,mode='circle',main="circle")
gplot(Bali,gmode="graph",edge.col="grey75",
      vertex.cex=1.5,mode='eigen',main="eigen")
gplot(Bali,gmode="graph",edge.col="grey75",
      vertex.cex=1.5,mode='random',main="random")
gplot(Bali,gmode="graph",edge.col="grey75",
      vertex.cex=1.5,mode='spring',main="spring")
gplot(Bali,gmode="graph",edge.col="grey75",
      vertex.cex=1.5,mode='fruchtermanreingold',
```

```
        main='fruchtermanreingold')
gplot(Bali,gmode="graph",edge.col="grey75",
        vertex.cex=1.5,mode='kamadakawai',
        main='kamadakawai')
par(op)
```

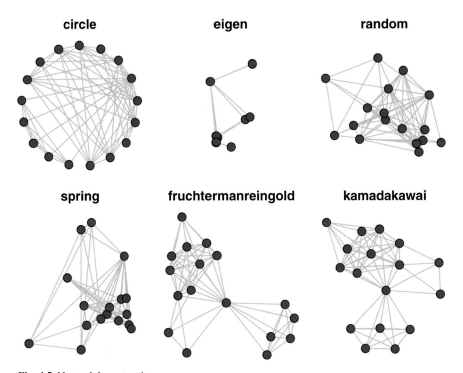

Fig. 4.5 Network layout options

4.3.1 Finer Control Over Network Layout

The layout options provided in statnet (and igraph, see below) work algorithmically or heuristically, usually with some randomness. So, even with the same layout option, a different graphic layout will be produced each time the network is plotted. Fortunately, R provides a way to have exact control over the layout coordinates. This allows for exact positioning, or saving the layout coordinates after a particular network is plotted.

The coord option in the plot function is used for this. This option expects a matrix with two columns. Each row corresponds to one node, the first column gives the X coordinate, and the second column gives the Y coordinate. Also, the results of a plotting function can be saved to an object, which will contain the coordinates of the produced plot. This is demonstrated in the next example. Here, we produce an initial plot of the Bali network, saving the coordinates. We then stretch out the layout of the graph by multiplying the Y coordinates by a constant. Both plots are shown in the figure, along with axes to make it easier to see how the coordinates have changed (Fig. 4.6). There are many other ways to use specific coordinates, but the main use is to preserve a particular layout for future production and examination.

```
mycoords1 <- gplot(Bali,gmode="graph",
        vertex.cex=1.5)
mycoords2 <- mycoords1
mycoords2[,2] <- mycoords1[,2]*1.5
mycoords1
```

```
##              x      y
##    [1,]  -6.299  11.84
##    [2,]  -3.887  13.80
##    [3,]  -8.355   9.89
##    [4,]  -4.672  10.28
##    [5,]  -8.537  11.65
##    [6,]  -5.932  12.63
##    [7,]  -2.420  12.96
##    [8,]  -7.694  10.09
##    [9,]  -8.334  10.86
##   [10,]  -0.935   8.08
##   [11,]  -3.015   6.98
##   [12,]  -1.863   7.10
##   [13,]  -1.094   9.30
##   [14,]  -2.061   8.51
##   [15,]  -7.715  12.83
##   [16,] -10.453  11.25
##   [17,]  -8.357  12.43
```

```
mycoords2
```

```
##              x      y
##    [1,]  -6.299  17.8
##    [2,]  -3.887  20.7
##    [3,]  -8.355  14.8
##    [4,]  -4.672  15.4
##    [5,]  -8.537  17.5
##    [6,]  -5.932  18.9
##    [7,]  -2.420  19.4
##    [8,]  -7.694  15.1
##    [9,]  -8.334  16.3
##   [10,]  -0.935  12.1
```

```
## [11,]   -3.015 10.5
## [12,]   -1.863 10.7
## [13,]   -1.094 14.0
## [14,]   -2.061 12.8
## [15,]   -7.715 19.2
## [16,] -10.453 16.9
## [17,]   -8.357 18.6
```

```
op <- par(mar=c(4,3,4,3),mfrow=c(1,2))
gplot(Bali,gmode="graph",coord=mycoords1,
      vertex.cex=1.5,suppress.axes = FALSE,
      ylim=c(min(mycoords2[,2])-1,max(mycoords2[,2])+1),
      main="Original coordinates")
gplot(Bali,gmode="graph",coord=mycoords2,
      vertex.cex=1.5,suppress.axes = FALSE,
      ylim=c(min(mycoords2[,2])-1,max(mycoords2[,2])+1),
      main="Modified coordinates")
par(op)
```

4.3.2 Network Graph Layouts Using `igraph`

The `igraph` package provides the user with a similar set of options for controlling
the layouts of network graphics. The `layout` option is used to specify an existing
layout function or refer to a set of vertex coordinates. See `?igraph.plotting`
for more information on plotting and layout options in `igraph` (Fig. 4.7).

```
detach(package:statnet)
library(igraph)
library(intergraph)
iBali <- asIgraph(Bali)
op <- par(mar=c(0,0,3,0),mfrow=c(1,3))
plot(iBali,layout=layout_in_circle,
              main="Circle")
plot(iBali,layout=layout_randomly,
              main="Random")
plot(iBali,layout=layout_with_kk,
              main="Kamada-Kawai")
par(op)
```

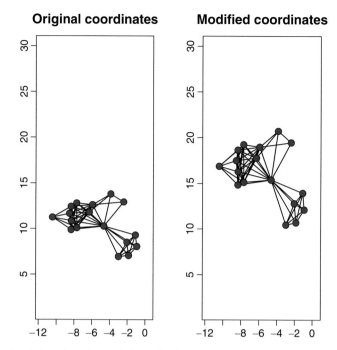

Fig. 4.6 Network layouts with modified coordinates

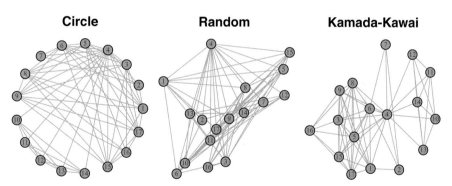

Fig. 4.7 Network layout options with igraph

Chapter 5
Effective Network Graphic Design

> As with any graphic, networks are used in order to discover
> pertinent groups or to inform others of the groups and structures
> discovered. It is a good means of displaying structures.
> However, it ceases to be a means of discovery when the elements
> are numerous. The figure rapidly becomes complex, illegible
> and untransformable. (Jacques Bertin)

5.1 Basic Principles

Achieving effective network graphic design is not that different from any other type
of information graphic. As Edward Tufte pointed out in his seminal *The Visual Display
of Quantitative Information,* "Graphical excellence is that which gives to the
viewer the greatest number of ideas in the shortest time with the least ink in the
smallest space." Network graphics actually start out with an important advantage in
that they typically have a high information/ink ratio.

The goal for any network graphic design should be to produce a figure that reveals
the important or interesting information that is contained in the network data. To
do this, the analyst must make decisions about every graphical element that can
appear in the figure. R, and the plotting functions contained in the `statnet` and
`igraph` packages, give the analyst almost complete programmatic control over the
appearance of the network graphic. The purpose of this chapter is to walk through
many of the most useful design elements in network graphics, and discuss how to
use them and why they should be used in certain ways.

5.2 Design Elements

Like any other type of information graphic, network visualizations are made up
of a large number of distinct visual elements. These individual elements include
things that are distinctive to network graphics, such as nodes and ties, as well as
other elements common to most graphics, such as titles, legends, etc. The plotting
functions in `statnet` and `igraph` provide a great deal of programmatic control
to the user.

Although a simple call to a plotting function is enough to produce a default network graphic, it is almost always the case that you will need to take time to set
appropriate function options and develop some additional R code to produce an

© Springer International Publishing Switzerland 2015
D.A. Luke, *A User's Guide to Network Analysis in R*, Use R!,
DOI 10.1007/978-3-319-23883-8_5

effective graphic. Some design decisions will be made on aesthetic grounds, while many others will be based on the most important pattern or story that you wish to convey with the graphic and that is supported by the underlying network data.

The following sections present a quick tour of the most commonly used individual graphing elements for network visualizations. They will each be covered on their own in turn.

5.2.1 Node Color

By default, `statnet` produces a network graphic with red circles as nodes. To designate a different color, the `vertex.col` option is used in the `gplot` function. (The `gmode` option is also used here to tell `statnet` not to handle Bali as a directed graph.) So, for example, it is a simple matter to produce a plot with attractive light blue nodes (Fig. 5.1).

```
data(Bali)
gplot(Bali,vertex.col="slateblue2",gmode="graph")
```

In general, all of the basic color-handling options of R are available for plotting networks. This opens up a lot of power and flexibility for graphic design, but to use color effectively will require some homework. In particular, it will be useful to read more in-depth treatments of color use in R (e.g., Murrell 2005).

As the above example suggests, a color can be designated by its color name. To see all of the 657 possible color names recognized by R, use the `colors()` command. In addition to specifying the name, colors can be selected using Red-Green-Blue (RGB) triplets of intensities. Alternatively, the RGB specification can also be provided using a hexadecimal string of the form '#RRGGBB', where each of the RR, GG, and BB parts of the string is a hexadecimal number that provides the red, green or blue intensity ranging from 00 to FF.

The following code will produce the same network graphic with the same light blue nodes (figure not shown), showing how you can obtain colors using the rgb and hexadecimal approaches. To get the appropriate rgb values for a particular color name, you can use the `col2rgb()` function. The hexadecimal codes were obtained at http://www.javascripter.net/faq/rgbtohex.htm.

```
col2rgb('slateblue2')
gplot(Bali,vertex.col=rgb(122,103,238,
                maxColorValue=255),gmode="graph")
gplot(Bali,vertex.col="#7A67EE",gmode="graph")
```

One less common color feature in R can come in handy for network diagrams, especially with large networks where the nodes overlap in the graphic. Normally, colors are fully opaque, so overlapping nodes in a graphic will lead to large color 'blobs' where it is hard to distinguish the nodes. However, it is possible to make

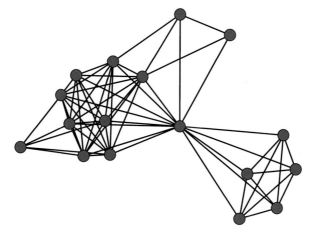

Fig. 5.1 Bali network with *light blue* nodes

use of partially transparent colors, using a built-in alpha transparency channel. The
rgb() function can be used to specify the amount of transparency, from 0 (fully
transparent) to 1 (fully opaque). See ?rgb for more details.

Figure 5.2 shows the difference when using the alpha transparency color channel.
Both graphics show the same random network of 300 nodes. (The layouts are different because each is using the Fruchterman-Reingold force-directed algorithm.) The
figure on the left is using a fully opaque dark blue color. The figure on the right is
still using dark blue, but with an alpha transparency channel of approximately 30 %.
The overlapping nodes are much easier to see when transparent colors are used.
Note that some graphics devices in R may not support transparent colors.

```
ndum <- rgraph(300,tprob=0.025,mode="graph")
op <- par(mar = c(0,0,2,0),mfrow=c(1,2))
gplot(ndum,gmode="graph",vertex.cex=2,
      vertex.col=rgb(0,0,139,maxColorValue=255),
      edge.col="grey80",edge.lwd=0.5,
      main="Fully opaque")
gplot(ndum,gmode="graph",vertex.cex=2,
      vertex.col=rgb(0,0,139,alpha=80,
                     maxColorValue=255),
      edge.col="grey80",edge.lwd=0.5,
      main="Partly transparent")
par(op)
```

In these previous examples, every node has the same color. A more important
use of color is to communicate some characteristic of the node or network by having
different nodes have different colors. Specifically, information stored in a categorical
node attribute can often be communicated through judicious node color choices.

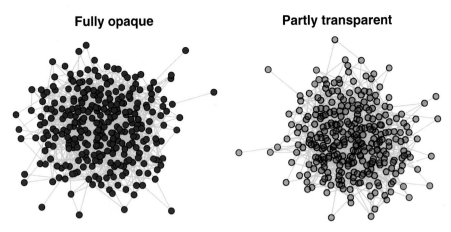

Fig. 5.2 Alpha transparency channel example

For example, the Bali terrorist network has the `role` vertex attribute which stores the categorical description of the role that each member played in the network. `CT` means a member of the command team, `BM` is a bomb maker, etc. (See `?Bali` for more information.) Node colors can be used to effectively distinguish the network member roles. Since this information is already stored in a vertex attribute, `statnet` can use this to automatically pick node colors. (This is only true for `plot()`, not `gplot()`.)

```
rolelab <- get.vertex.attribute(Bali,"role")
op <- par(mar=c(0,0,0,0))
plot(Bali,usearrows=FALSE,vertex.cex=1.5,label=rolelab,
     displaylabels=T,vertex.col="role")
par(op)
```

Figure 5.3 displays the Bali network with nodes colored according to the role each member played in the network. (The node labels are also printed out to facilitate interpretation. Node labeling will be discussed in Sect. 5.2.4.) The network is much more interpretable by using color coding in this way. For example, we can more easily understand the subgroup structure by noting that the greater density between the members of Team Lima (TL, cyan), as well as the bombmakers (BM, black).

However, by simply using the name of an existing vertex attribute, `statnet` picks node colors from the existing default color palette in R. Viewing this palette, we can see that "BM" is assigned to the color black because "BM" comes first alphabetically in the role attribute string, and black is the first entry in the color palette. "CT" comes second, so it is assigned red which is the 2nd entry in the palette, an so on.

```
palette()
```

```
## [1] "black"    "red"      "green3"  "blue"
## [5] "cyan"     "magenta" "yellow"  "gray"
```

Using the default palette has a number of disadvantages. First, it is limited to eight colors. (R will cycle through the set of eight colors if there are more than eight types of nodes to color.) Second, the default palette starts with black which is often not a good color choice to include with other colors in a network graphic. More generally, the default colors do not represent an aesthetically pleasing or useful set of colors for displaying categorical classifications.

A more flexible, and usually aesthetically more satisfying approach is to set up your own color palette and then index into it for color selection. The `RColor Brewer` package provides a number of predesigned color palettes that are very useful when using color to distinguish between a relatively small set of categories. For more information, see `?RColorBrewer`. More details on the ColorBrewer

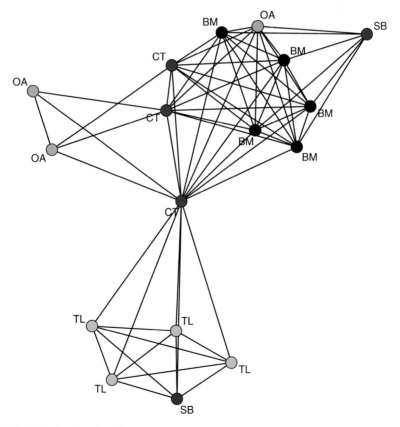

Fig. 5.3 Bali network colored by role

system are available at `http://www.colorbrewer.org`. (There are many other color and palette picking options available, for example see the interactive palette chooser at `http://paletton.com`.)

```
library(RColorBrewer)
display.brewer.pal(5, "Dark2")
```

In the following code, a user-defined palette is created by selecting five colors from a larger palette called `Dark2` provided by `RColorBrewer`. Once the palette has been defined, it can be used in the network plotting call. This approach produces more pleasing sets of colors, and is much more flexible than relying on the default color palette (Fig. 5.4).

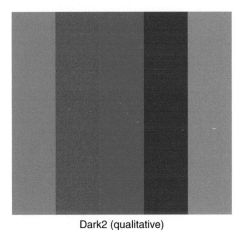

Dark2 (qualitative)

Fig. 5.4 A set of five colors chosen from an RColorBrewer palette

Note that we convert the `role` vertex attribute character vector to a factor so that the indexing will work. This means that if you have a numeric vector stored as a vertex attribute that you do not have to turn it into a factor. In other words, the indexing works with factors or numeric vectors, but not character vectors (Fig. 5.5).

```
my_pal <- brewer.pal(5,"Dark2")
rolecat <- as.factor(get.vertex.attribute(Bali,"role"))
plot(Bali,vertex.cex=1.5,label=rolelab,
     displaylabels=T,vertex.col=my_pal[rolecat])
```

5.2.2 Node Shape

In addition to using color to distinguish between different types of nodes, `statnet` can be directed to use different shapes for the nodes. This is mainly useful when

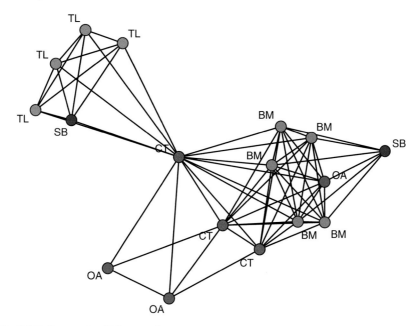

Fig. 5.5 Bali network with better colors

there is a small number of node types. It is also particularly useful for situations where you will not be able to use color to distinguish nodes (or help viewers who may be color-blind).

Unfortunately, statnet has only a limited ability to distinguish nodes by shapes, by designating the number of sides used to plot the node polygon (normally, the number of sides is 50, which produces a circle). If the number of sides is 3 you get a triangle, 4 a square, and so on. This is only useful for a very small number of node types (Fig. 5.6).

If you have a particular need to use node shapes in a network graphic, igraph is much more flexible in this regard. See Sect. 9.2.3 for an igraph plotting example with different node shapes.

```
op <- par(mar=c(0,0,0,0))
sidenum <- 3:7
plot(Bali,usearrows=FALSE,vertex.cex=4,
    displaylabels=F,vertex.sides=sidenum[rolecat])
par(op)
```

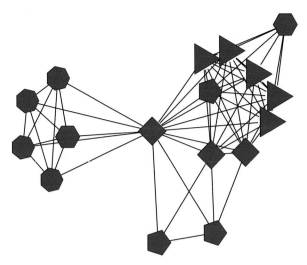

Fig. 5.6 Bali network with different node shapes

5.2.3 Node Size

Network node sizes are controlled by the `vertex.cex` option in the `statnet` `plot` and `gplot` functions (similar to how sizes of graphic elements are controlled in the base *R* graphics system). The overall sizes of the nodes should be set so that the nodes are large enough to be distinguishable, but small enough that they do not extensively overlap. In the following 'Goldilocks' example, we can see how `vertex.cex` can be adjusted to find an effective node size (Fig. 5.7).

```
op <- par(mar = c(0,0,2,0),mfrow=c(1,3))
plot(Bali,vertex.cex=0.5,main="Too small")
plot(Bali,vertex.cex=2,main="Just right")
plot(Bali,vertex.cex=6,main="Too large")
par(op)
```

Rather than setting the same overall size for every node, it is often useful to use the node size in a network graphic to communicate some important quantitative characteristic. For example, nodes vary in their positions in the overall network. Some nodes are very central, while others are more peripheral. Chapter 7 discusses node prominence and centrality in more detail, but for now we will simply calculate some node characteristics such that larger numbers indicate more central nodes.

To set this up, we will calculate three different measures of node centrality. Each of these lines of code produces a vector of centrality measures for each node, and larger numbers indicate greater centrality.

```
deg <- degree(Bali,gmode="graph")
deg
```

```
## [1]   9   4   9  15   9  10   3   9   9   5   5   5   5   5   9
## [16]  6   9
```

```
cls <- closeness(Bali,gmode="graph")
cls
```

```
## [1] 0.696 0.552 0.696 0.941 0.696 0.727 0.533
## [8] 0.696 0.696 0.571 0.571 0.571 0.571 0.571
## [15] 0.696 0.485 0.696
```

```
bet <- betweenness(Bali,gmode="graph")
bet
```

```
## [1]   2.333   0.333   1.667  61.167   1.667   6.167
## [7]   0.000   1.667   1.667   0.000   0.000   0.000
## [13]  0.000   0.000   1.667   0.000   1.667
```

Once you have this node-level vector of quantitative information, it can be used to set the relative sizes of the nodes. This is done by using the same vertex.cex

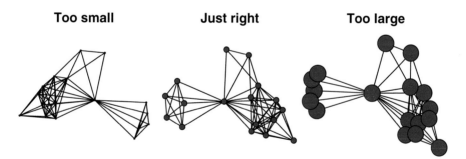

Fig. 5.7 Adjusting overall node size

option as before, but instead of assigning a single number we assign the vector of node information.

```
op <- par(mar = c(0,0,2,1),mfrow=c(1,2))
plot(Bali,usearrows=T,vertex.cex=deg,main="Raw")
plot(Bali,usearrows=FALSE,vertex.cex=log(deg),
     main="Adjusted")
par(op)
```

However, as we can see by comparing the two panels in Fig. 5.8, the raw numbers in the deg vector produce nodes that are much too large. They need to be adjusted, and in this case we get usable sizes by taking the log of the deg values. This results in node sizes where we can more easily see the nodes with higher degree relative to nodes with lower degree.

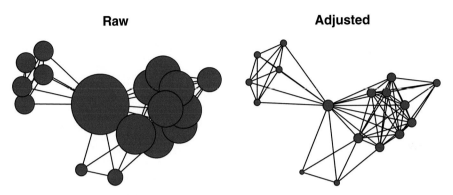

Fig. 5.8 Adjusting relative node size – Example 1

The next two examples show other types of adjustments that might be necessary when setting relative node sizes. Using cls (closeness) we have the opposite problem from the previous example, where the nodes sizes start out too small. So an appropriate adjustment is to multiply the original values (Fig. 5.9). The bet vector (betweenness) provides a more complex example. First, the raw vector sizes vary across several orders of magnitude (with one node with a size of 122.3). In addition, some of the nodes have 0 for their bet values. These zeros would result in the nodes being plotted with 0 size, so we need to handle this by adding 1 to the entire vector before taking the square root (Fig. 5.10) .

```
op <- par(mar = c(0,0,2,1),mfrow=c(1,2))
plot(Bali,usearrows=T,vertex.cex=cls,main="Raw")
plot(Bali,usearrows=FALSE,vertex.cex=4*cls,
     main="Adjusted")
par(op)
```

```
op <- par(mar = c(0,0,2,1),mfrow=c(1,2))
plot(Bali,usearrows=T,vertex.cex=bet,main="Raw")
plot(Bali,usearrows=FALSE,vertex.cex=sqrt(bet+1),
     main="Adjusted")
par(op)
```

The adjustments for relative node sizes can be tedious, although R does give you complete control for how to adjust the sizes. The following function can be used to save some time when figuring out the best node sizes. The function rescale() takes a vector of node characteristics (actually can be any numeric vector), and rescales the values to fit between the low and high values.

Raw   **Adjusted**

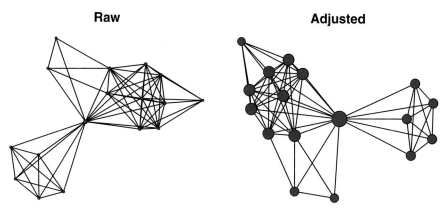

Fig. 5.9 Adjusting relative node size – Example 2

Raw   **Adjusted**

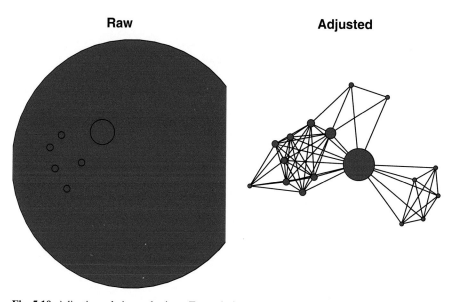

Fig. 5.10 Adjusting relative node size – Example 3

```
rescale <- function(nchar,low,high) {
  min_d <- min(nchar)
  max_d <- max(nchar)
  rscl <- ((high-low)*(nchar-min_d))/(max_d-min_d)+low
  rscl
}
```

The next plot shows how the function works and rescales the raw degree values
for the Bali network to set the node sizes to vary from one to six (Fig. 5.11).

```
plot(Bali,vertex.cex=rescale(deg,1,6),
     main="Adjusted node sizes with rescale function.")
```

Adjusted node sizes with rescale function.

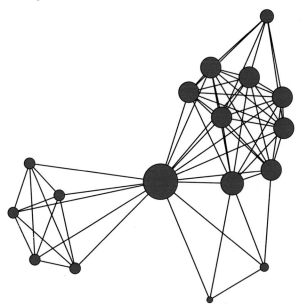

Fig. 5.11 Rescaling the node size based on degree

5.2.4 Node Label

A network graphic is often more interesting and easier to interpret if nodes are labelled so that the audience can see who or what makes up the network. This is particularly helpful for smaller networks; if networks get too large then the labels themselves may get in the way of the network information.

   If a network object in statnet contains the special vertex attribute
vertex.names, then this can be used to automatically display node labels when plotting. Other characteristics of the node labels can be controlled such as font size, color, and distance from node (Fig. 5.12).

```
get.vertex.attribute(Bali,"vertex.names")

   ##  [1] "Muklas"   "Amrozi"   "Imron"    "Samudra"
   ##  [5] "Dulmatin" "Idris"    "Mubarok"  "Husin"
```

```
##   [9]  "Ghoni"      "Arnasan"   "Rauf"      "Octavia"
##  [13]  "Hidayat"    "Junaedi"   "Patek"     "Feri"
##  [17]  "Sarijo"
```

```
op <- par(mar = c(0,0,0,0))
plot(Bali,displaylabels=TRUE,label.cex=0.8,
     pad=0.4,label.col="darkblue")
par(op)
```

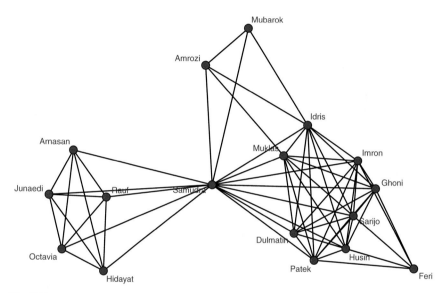

Fig. 5.12 Bali network with labelled nodes.

The automatic labels based on information stored in the `vertex.names` attribute may not be the most important or useful information. For example, in the case of the Bali network the actual names of the terrorists are not that interesting to most viewers. Fortunately, you can use other text information to label the nodes. We saw an example of this earlier in Fig. 5.3. In this case we are using the text stored in the `role` vertex attribute to label the nodes. The key here is to use the `label` option to specify what text vector to use for the labels (Fig. 5.13).

```
rolelab <- get.vertex.attribute(Bali,"role")
plot(Bali,usearrows=FALSE,label=rolelab,
     displaylabels=T,label.col="darkblue")
```

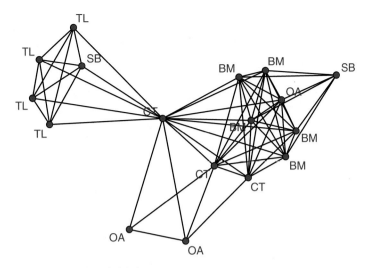

Fig. 5.13 Bali network with role labels

5.2.5 Edge Width

If your network data include valued ties, or in general any quantitative information that can be related to ties between nodes, then you can communicate that information visually by altering the width of the displayed ties in a network graphic. For example, the strength of friendship ties might be known, or the amount of money that flows between organizations in a directed network might be measured. In these cases, thicker ties can denote greater strength or greater flow (Fig. 5.14).

The Bali network includes a tie attribute called IC, which is a simple five-level ordinal scale that was used to measure the amount of interaction between members of the network. This attribute can be used to set the width of the ties in the network visualization. In the example below the IC values are extracted from the stored edge attribute, this allows us to transform the vector to better distinguish among the five IC levels (by multiplying the vector by 1.5).

```
op <  par(mar = c(0,0,0,0))
IClevel <- Bali %e% "IC"
plot(Bali,vertex.cex=1.5,
     edge.lwd=1.5*IClevel)
par(op)
```

5.2.6 Edge Color

While edge width can be set to communicate quantitative information about network ties, the color of the edge can be set to communicate qualitative information about the tie, similar to how node colors can set. For example, you could use different colors of line graphics to distinguish between positive and negative ties in a social network (Fig. 5.15).

The Bali network does not contain categorical or qualitative information stored in an edge attribute, so here we create a random categorical vector to demonstrate how to use different edge colors in a network graphic. For this example, we set up a color palette that can be used to index the correct color choice, based on the categorical edge vector. In this case blue will be used for edge type #1, red for edge type #2, and green for edge type #3. This might reflect neutral ties (blue), negative ties (red), and positive ties (green). (Also see Fig. 7.8 in Chap. 7 for a more realistic example of using different line colors.)

```
n_edge <- network.edgecount(Bali)
edge_cat <- sample(1:3,n_edge,replace=T)
linecol_pal <- c("blue","red","green")
```

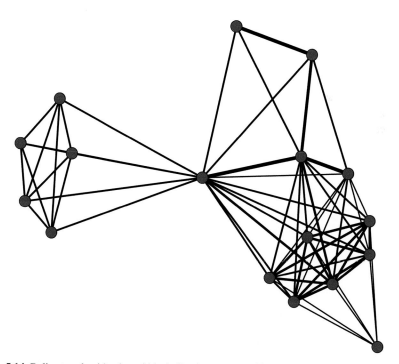

Fig. 5.14 Bali network with edge widths indicating amount of interaction

```
plot(Bali,vertex.cex=1.5,vertex.col="grey25",
     edge.col=linecol_pal[edge_cat],edge.lwd=2)
```

5.2.7 Edge Type

While edge width can be set to communicate quantitative information about network ties, the type of the edge can be set to communicate qualitative information about the ties. For example, you could use different types of line graphics to distinguish between positive and negative ties in a social network.

The Bali network does not contain categorical or qualitative information stored in an edge attribute, so here we create a random categorical vector to demonstrate how to use different edge types in a network graphic. Here three different line types are used (2 = dashed; 3 = dotted; 4 = dotdash). Also, the different line types do not show up clearly using plot(), so gplot() is used here (Fig. 5.16).

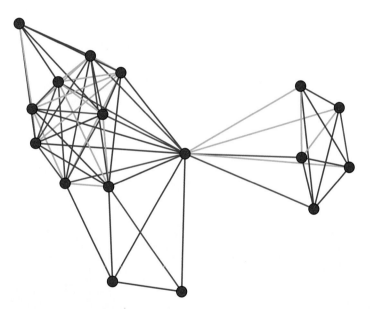

Fig. 5.15 Bali network with different edge colors

```
n_edge <- network.edgecount(Bali)
edge_cat <- sample(1:3,n_edge,replace=T)
line_pal <- c(2,3,4)
```

```
gplot(Bali,vertex.cex=0.8,gmode="graph",
      vertex.col="gray50",edge.lwd=1.5,
      edge.lty=line_pal[edge_cat])
```

Although this works as intended, the resulting graphic is not very attractive and (in my mind) is hard to interpret. Different line types should be used sparingly, and probably only for very small networks with only two different line types. Most published network graphics stick to color and maybe line width to distinguish among different types of network ties.

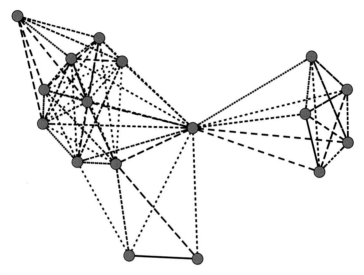

Fig. 5.16 Bali network with different edge types

5.2.8 Legends

The examples above show how network graphic elements such as node color, node shape, node size, edge type, edge width can be used to communicate important characteristics of the network. As with other types of information graphics, it is often useful to provide a legend so that the meaning of this information is clear to the user.

The basic plotting functions contained in statnet do not have built-in functionality for providing a network graphic legend. Fortunately, it is easy to use the legend() function provided by basic R to add a legend to a network graphic. In the example below we replicate the network graphic from Fig. 5.5 but add a legend to provide the node color key. We also scale the node sizes to reflect node

prominence based on degree. See ?legend for more details on how to use legends.
Figure 5.17, in fact, serves as a nice example of a carefully designed network graphic
that could be used as a final product. It uses node size, node color, and a legend to
efficiently and clearly communicate the most important information contained in
the Bali network.

```
my_pal <- brewer.pal(5,"Dark2")
rolecat <- as.factor(get.vertex.attribute(Bali,"role"))
plot(Bali,vertex.cex=rescale(deg,1,5),
     vertex.col=my_pal[rolecat])
legend("bottomleft",legend=c("BM","CT","OA","SB","TL"),
       col=my_pal,pch=19,pt.cex=1.5,bty="n",
       title="Terrorist Role")
```

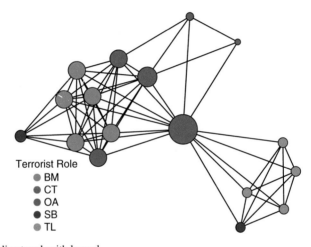

Fig. 5.17 Bali network with legend

Chapter 6
Advanced Network Graphics

One eye sees, the other feels. (Paul Klee)

As the previous two chapters demonstrate, both `statnet` and `igraph` have sophisticated plotting capabilities that can produce a very wide variety of network graphics. However, these plotting functions cannot meet all of the analytic or presentation needs. In particular, network scientists may wish to produce more specialized network graphics. Also, while `statnet` and `igraph` excel at producing high-quality publication ready network graphics, these graphics are static. Fortunately, developers have started exploring how to take network graphics and deliver them to web-based platforms where users can interact with the diagrams. This chapter explores a few of these more specialized network graphic techniques, as well as demonstrating how to produce some simple web-based interactive network diagrams.

6.1 Interactive Network Graphics

One of the useful features of many other network analysis packages such as UCINet and Pajek is the ability to produce network diagrams that are interactive at some level. For example, in Pajek a network visualization can be produced in a separate 'Draw' window, and then the user can interact with that window in various ways to edit or change the network graphic. These capabilities can be very useful for exploring the network, as well as fine-tuning a network graphic for subsequent dissemination.

Although R's programmatic framework allows for detailed control over all the elements of a network graphic, this is generally not made available to the user in an interactive way. There are a few exceptions to this, as well as some new packages that allow for creating interactive network diagrams that can be published to the web. In this section a few of these options are demonstrated.

© Springer International Publishing Switzerland 2015

D.A. Luke, *A User's Guide to Network Analysis in R*, Use R!,

DOI 10.1007/978 3 319 23883 8_6

6.1.1 Simple Interactive Networks in `igraph`

The `igraph` package includes the `tkplot()` function which supports simple
interactive network plots through a Tk graphics window. Only some features of
the network graphics can be modified. A typical use for this feature is to produce the
interactive graphic, adjust the node positions to improve the network layout, save the
node position coordinates and then use the coordinates to produce a final (non-
interactive) network diagram. This work flow is illustrated below with the `Bali`
network, see Chap. 8 for a more in-depth example with these data.

```
library(intergraph)
library(igraph)
data(Bali)
iBali <- asIgraph(Bali)
Coord <- tkplot(iBali, vertex.size=3,
                vertex.label=V(iBali)$role,
                vertex.color="darkgreen")
# Edit plot in Tk graphics window before
# running next two commands.
MCoords <- tkplot.getcoords(Coord)
plot(iBali, layout=MCoords, vertex.size=5,
          vertex.label=NA, vertex.color="lightblue")
```

6.1.2 Publishing Web-Based Interactive Network Diagrams

Instead of building interactive network graphics within R itself, more people are
beginning to look at ways to produce interactive graphics that are published on the
Web, using frameworks like the D3 JavaScript library (http://http://d3js.
org/) and Shiny (http://shiny.rstudio.com/).

None of these approaches yet have come close to matching what a fully-developed
network graphics application such as Gephi can do. However, I anticipate that we
will be seeing rapid development of more R-connected approaches to web-based
network visualization in the next few years.

The `networkD3` package is a small set of functions that can be used to build
simple interactive network graphics that can be displayed in shiny-aware documents
(i.e., RStudio) or in HTML web-pages. The following code shows how simple it is to
produce an interactive graphic. The first set of lines will send a graphic to the Viewer
window if you run the commands within RStudio. The `simpleNetwork()`
function expects the network data in the form of an edgelist stored in a dataframe.
(The output from the examples in this section is not shown here, because it requires
RStudio or a web browser to view.)

```
library(networkD3)
src <- c("A","A","B","B","C","E")
target <- c("B","C","C","D","B","C")
net_edge <- data.frame(src, target)
simpleNetwork(net_edge)
```

To save the interactive network to a freestanding HTML file, use the following code.

```
net_D3 <- simpleNetwork(net_edge)
saveNetwork(net_D3,file = 'Net_test1.html',
            selfcontained=TRUE)
```

The output from simpleNetwork is so simple that it mainly is useful as a proof-of-concept or tech demo. Slightly more sophisticated network graphics can be produced using the forceNetwork() function. For this example, we are using the Bali network again. The function expects data to be passed to it in two data frames. The 'links' dataframe will have the network data in edgelist format. The 'nodes' dataframe will have the node id and properties of the nodes. Currently only a categorical grouping variable is allowed. If the nodes have numeric ids, they must start at 0. So, the main work to use the function is putting the data into the correct format.

```
iBali_edge <- get.edgelist(iBali)
iBali_edge <- iBali_edge - 1
iBali_edge <- data.frame(iBali_edge)
iBali_nodes <- data.frame(NodeID=as.numeric(V(iBali)-1),
                          Group=V(iBali)$role,
                          Nodesize=(degree(iBali)))
forceNetwork(Links = iBali_edge, Nodes = iBali_nodes,
             Source = "X1", Target = "X2",
             NodeID = "NodeID",Nodesize = "Nodesize",
             radiusCalculation="Math.sqrt(d.nodesize)*3",
             Group = "Group", opacity = 0.8,
             legend=TRUE)
```

Once again, this can be saved to an external file. Be careful, you will get an error if you try to overwrite an existing file, even if it is not open in your browser.

```
net_D3 <- forceNetwork(Links = iBali_edge,
             Nodes = iBali_nodes,
             Source = "X1", Target = "X2",
             NodeID = "NodeID",Nodesize = "Nodesize",
             radiusCalculation="Math.sqrt(d.nodesize)*3",
             Group = "Group", opacity = 0.8,
             legend=TRUE)
```

```
saveNetwork(net_D3,file = 'Net_test2.html',
            selfcontained=TRUE)
```

The `visNetwork` package is a similar set of tools that uses the `vis.js` javascript library (`http://visjs.org/`) to produce web-based interactive network graphics.

This package also requires network data to be provided in a nodes data frame and an edges data frame. The nodes data frame should include an `id` column, and the edges data frame should have `from` and `columns`. Using the Bali network, the following code sets up the data and produces a minimal example of an interactive network graphic. Like in the previous example, this code produces an interactive network in the Viewer window of RStudio.

```
library(visNetwork)
iBali_edge <- get.edgelist(iBali)
iBali_edge <- data.frame(from = iBali_edge[,1],
                          to = iBali_edge[,2])
iBali_nodes <- data.frame(id = as.numeric(V(iBali)))
visNetwork(iBali_nodes, iBali_edge, width = "100%")
```

The `visNetwork` package has a large number of options that can be used to control the appearance of the network diagram, as well as for controlling how the plot can be embedded in `Shiny` web applications. See the package help file for more information, as well as a more in-depth demonstration of its capabilities available at `http://dataknowledge.github.io/visNetwork/`. The next code shows off some of these options.

```
iBali_nodes$group <- V(iBali)$role
iBali_nodes$value <- degree(iBali)
net <- visNetwork(iBali_nodes, iBali_edge,
                  width = "100%",legend=TRUE)
visOptions(net,highlightNearest = TRUE)
```

First, some of the display options are controlled by saving node or edge information into the `nodes` or `edges` data frames. Here, the `group` variable stores the 'role' attribute, and the `value` variable is used to store the node sizes (in this case, the degree). The `visNetwork()` and `visOptions()` functions are used to display the network, add a legend based on the grouping variable, set default colors for each group, and then allow for the user to highlight individual nodes and their immediate neighbors when clicking on a node in the diagram.

As before, these interactive plots will appear in a plot window if you are using RStudio. Once the plot has been designed, it can be exported to a freestanding webpage or embedded in other web platforms (e.g., with `Shiny`). This last example shows how to save the plot in a separate web file, using the `saveWidget()` function from the `htmlwidgets` package, which is installed when you install the

`visNetwork` package. This example adds a set of navigation buttons to the final
network plot that allows moving the network and zooming in or out.

```
net <- visNetwork(iBali_nodes, iBali_edge,
                  width = "100%",legend=TRUE)
net <- visOptions(net,highlightNearest = TRUE)
net <- visInteraction(net,navigationButtons = TRUE)
library(htmlwidgets)
saveWidget(net, "Net_test3.html")
```

6.1.3 Statnet Web: Interactive `statnet` with `shiny`

As evidence of the rapid development of interactive network tools, the Statnet development team has recently published a web-based version of their R network analytic
tools using the `shiny` web application framework.

   Statnet Web can be used by connecting directly to the *shinyapps.io* server
at `https://statnet.shinyapps.io/statnetWeb`. Or, the tools can be
run locally by installing the `statnetWeb` package. In addition to producing
basic network plots by selecting parameters and options from drop-down boxes,
`statnetWeb` can produce a variety of network statistics as well as fit and test
ERGMs (see Chap. 11). Although web-based statnet does not give as much control
over or reproducibility of network analytic results as a programming approach does,
it is an impressive platform for quickly exploring network characteristics and will
be useful for teaching as well as disseminating network analytic results.

```
library(statnetWeb)
run_sw()
```

6.2 Specialized Network Diagrams

Traditionally, network diagrams are plotted to illustrate fundamental network and
node properties such as prominence (see Chap. 4). However, there are a number
of more specialized plotting techniques that can be used that are appropriate for
highlighting other important or interesting aspects of the networks. Three of these
approaches are demonstrated in this section: arc diagrams, chord diagrams, and
heatmaps.

6.2.1 Arc Diagrams

Arc diagrams can be used when the positioning of nodes in a network is of less interest than the pattern of ties. Here is a simple example of an arc diagram, using the arcdiagram package. Note that this has to be installed using GitHub.

```
library(devtools)
install_github("gastonstat/arcdiagram")
```

The set-up for this example includes loading all the required libraries, then creating an edgelist object for the arcdiagram() function. For this example, we are using the Simpsons dataset, which contains a set of (fictitious) network data that shows the primary interaction ties between 15 of the characters on the Simpsons television show.

```
library(arcdiagram)
library(igraph)
library(intergraph)
data(Simpsons)
iSimp <- asIgraph(Simpsons)
simp_edge <- get.edgelist(iSimp)
```

A basic arc diagram can be produced with one function call (Fig. 6.1).

```
arcplot(simp_edge)
```

The arc diagram can be enhanced in a number of ways to highlight node and other network characteristics. Here we define some subgroups in the network (1 = family, 2 = work, 3 = school, 4 = neighborhood) and use colors to distinguish the groups (colors taken from a palette at colorbrewer2.org). Also, the degree of each node is used to adjust its size (Fig. 6.2).

```
s_grp <- V(iSimp)$group
s_col = c("#a6611a", "#dfc27d","#80cdc1","#018571")
cols = s_col[s_grp]
node_deg <- degree(iSimp)
```

```
arcplot(simp_edge, lwd.arcs=2, cex.nodes=node_deg/2,
        labels=V(iSimp)$vertex.names,
        col.labels="darkgreen",font=1,
        pch.nodes=21,line=1,col.nodes = cols,
        bg.nodes = cols, show.nodes = TRUE)
```

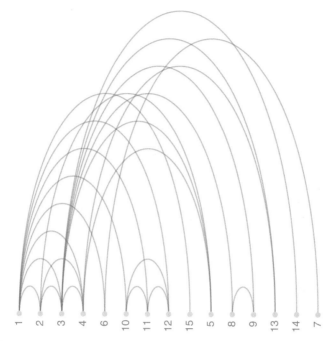

Fig. 6.1 Simpsons contact network

6.2.2 Chord Diagrams

Chord diagrams are a specialized type of information graphic that uses a circular layout to display the interrelationships between data in a matrix. They have become particularly popular in genetics research. Because network information can be organized in matrices, chord diagrams are an interesting graphic option for network plots. This is especially true for valued (weighted) and directed networks, where the amount and direction of the 'flows' are of interest.

The `circlize` package, by Zuguang Gu, implements a variety of circular graphics, including chord diagrams. The package has a lot of features, giving the user great control over the graphical appearance. The included vignette, `circular_visualization_of_matrix` is suggested reading.

In this example, we return to the network of the 2010 Netherlands World Cup soccer team. Although Fig. 1.2 shows the basic pattern of passing flows between the eleven members of the team, it ignores the number of passes (stored in the vertex attribute `passes`). Here we will create a chord diagram to further examine these patterns.

The first steps are to load the required packages and prepare the data. The main requirement is to have the network data in the form of a sociomatrix, with the entries corresponding to the strength or size of the tie if it is a valued network. The matrix will also have to have names assigned for the rows and columns. (In this example

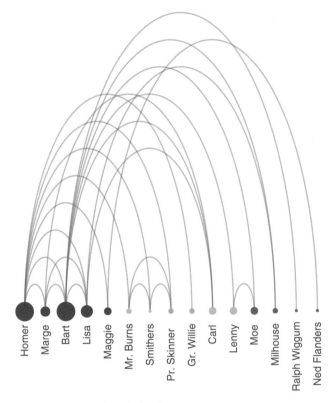

Fig. 6.2 Simpsons contact network – Version 2

we have an N×N matrix, so the names will be the same for rows and columns. The
circlize package can also be used for N×k matrices, so chord diagrams will
also be useful for 2-mode affiliation networks, such as those discussed in Chap. 9.)

```
library(statnet)
library(circlize)
data(FIFA_Nether)
FIFAm <- as.sociomatrix(FIFA_Nether,attrname='passes')
names <- c("GK1","DF3","DF4","DF5","MF6",
           "FW7","FW9","MF10","FW11","DF2","MF8")
rownames(FIFAm) = names
colnames(FIFAm) = names
FIFAm

   ##        GK1 DF3 DF4 DF5 MF6 FW7 FW9 MF10 FW11 DF2
   ## GK1      0  42  67  21   2  27   7    5    2  17
   ## DF3     30   0  44  14  42  15   8    7   10  36
   ## DF4     38  43   0  57  18  11   7   21    1   7
```

```
## DF5     6  14  47   0  11  50  20  40   1   4
## MF6     9  28  25  10   0  41  28  37  14  34
## FW7     4  12   1  21  21   0  15  33   9  25
## FW9     0   0   1   8   7  12   0  31  16   7
## MF10    1  11  11  22  43  29  20   0  28  13
## FW11    3   2   2   3   7   6  11  15   0  21
## DF2    29  38   8   3  45  38  10  18  26   0
## MF8    12  25  26  38  23  13  12  32  11  24
##        MF8
## GK1      3
## DF3     29
## DF4     28
## DF5     42
## MF6     21
## FW7     18
## FW9      2
## MF10    21
## FW11    12
## DF2     15
## MF8      0
```

The sociomatrix reveals a number of ties that have very low numbers of passes. To make the subsequent graphics a little easier to interpret we drop all ties with less than ten passes.

```
FIFAm[FIFAm < 10] <- 0
FIFAm
```

```
##        GK1 DF3 DF4 DF5 MF6 FW7 FW9 MF10 FW11 DF2
## GK1      0  42  67  21   0  27   0    0    0  17
## DF3     30   0  44  14  42  15   0    0   10  36
## DF4     38  43   0  57  18  11   0   21    0   0
## DF5      0  14  47   0  11  50  20   40    0   0
## MF6      0  28  25  10   0  41  28   37   14  34
## FW7      0  12   0  21  21   0  15   33    0  25
## FW9      0   0   0   0   0  12   0   31   16   0
## MF10     0  11  11  22  43  29  20    0   28  13
## FW11     0   0   0   0   0   0  11   15    0  21
## DF2     29  38   0   0  45  38  10   18   26   0
## MF8     12  25  26  38  23  13  12   32   11  24
##        MF8
## GK1      0
## DF3     29
## DF4     28
## DF5     42
## MF6     21
```

```
## FW7    18
## FW9     0
## MF10   21
## FW11   12
## DF2    15
## MF8     0
```

With a sociomatrix that has names assigned, a basic chord diagram can be produced by a simple call to the chordDiagram() function (Fig. 6.3).

```
chordDiagram(FIFAm)
```

Chord diagrams can contain a lot of information, especially for larger networks, so it is usually important to fine tune the plot to highlight the most important information. In this next plot, a number of options are used to make the graphic a little easier to interpret. First, colors are set so that players in the same position (Forward, Midfielder, etc.) have the same color. Then, because this is a directed network, flows (passes, in this case) go in both directions. The directional option is used so that the departing passes start further away from outer circle, making it easier to see the difference between passes sent and passes received. Finally, the order option is used to sort the players by their position.

```
grid.col <- c("#AA3939",rep("#AA6C39",4),
              rep("#2D882D",3),rep("#226666",3))
chordDiagram(FIFAm,directional = TRUE,
             grid.col = grid.col,
             order=c("GK1","DF2","DF3","DF4","DF5",
                     "MF6","MF8","MF10","FW7",
                     "FW9","FW11"))
```

In the resulting chord diagram (Fig. 6.4), it is much easier to see the patterns of passes among the players. We can see that FW7 receives more than twice the number of passes than the other two forwards. Similarly, we can see that the goalkeeper's favorite target is DF4, and that DF4 likes to pass frequently to DF5.

6.2.3 Heatmaps for Network Data

Heatmaps are another example of a specialized graphic that can be used for networks, especially valued or weighted networks. Here, a heatmap is produced to highlight the players who are passing or receiving the most among the Netherlands teammates.

First, a sociomatrix is created with the cells reflecting the tie weight, in this case 'passes.' Row and column names are defined for the margin labels.

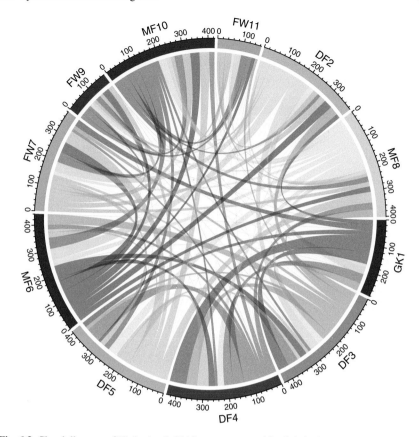

Fig. 6.3 Chord diagram of Netherlands 2010 soccer team, with all default options

```
data(FIFA_Nether)
FIFAm <- as.sociomatrix(FIFA_Nether,attrname='passes')
colnames(FIFAm) <- c("GK1","DF3","DF4","DF5",
                     "MF6","FW7","FW9","MF10",
                     "FW11","DF2","MF8")
rownames(FIFAm) <- c("GK1","DF3","DF4","DF5",
                     "MF6","FW7","FW9","MF10",
                     "FW11","DF2","MF8")
```

Once the data are set up, the heatmap is relatively easy to produce (Fig. 6.5). The colorRampPalette() function is used to designate a color range that will be used for the low and high ends of the values in the sociomatrix. (The color ranges chosen here were taken from color chooser tools at paletton.com.) The network data are directed, so it is important to remember that here the rows are the 'passers' and the columns are the 'receivers.'

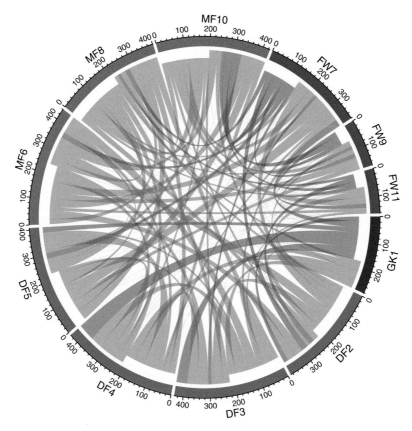

Fig. 6.4 Chord diagram of Netherlands 2010 soccer team, with advanced options

```
palf <- colorRampPalette(c("#669999", "#003333"))
heatmap(FIFAm[,11:1],Rowv = NA,Colv = NA,col = palf(60),
        scale="none", margins=c(11,11) )
```

The heatmap also shows the same pattern of heavy passers as Fig. 6.4. The darkest square is for the passes from the goalkeeper to DF4.

6.3 Creating Network Diagrams with Other R Packages

6.3.1 Network Diagrams with ggplot2

Although ggplot2 is not designed to handle all of the requirements of a full-fledged network visualization package, some of its advanced graphics capabilities can be used to create specialized network plotting routines. The following example

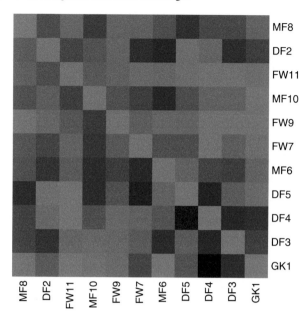

Fig. 6.5 Heatmap of Netherlands 2010 soccer team number of passes

is based on code developed by David Sparks, and posted on the blog he runs with his colleague Christopher DeSante, *is.R()* (http://is-r.tumblr.com).

The edgeMaker() function and supporting code can be used to create attractive and functional plots of directed networks using 'tapered-intensity-curved' edges. The bulk of the work is done by the edgeMaker() function which creates the curved ties between each connected dyad.

```
edgeMaker <- function(whichRow,len=100, curved = TRUE){
  fromC <- layoutCoordinates[adjacencyList[whichRow,1],]
  toC <- layoutCoordinates[adjacencyList[whichRow,2],]
  graphCenter <- colMeans(layoutCoordinates)
  bezierMid <- c(fromC[1], toC[2])
  distance1 <- sum((graphCenter - bezierMid)^2)
  if(distance1 < sum((graphCenter - c(toC[1],
                    fromC[2]))^2)){
    bezierMid <- c(toC[1], fromC[2])
    }
  bezierMid <- (fromC + toC + bezierMid) / 3
  if(curved == FALSE){bezierMid <- (fromC + toC) / 2}

  edge <- data.frame(bezier(c(fromC[1], bezierMid[1],
                    toC[1]),
                    c(fromC[2], bezierMid[2],
                    toC[2]),
                    evaluation = len))
```

```
edge$Sequence <- 1:len
edge$Group <- paste(adjacencyList[whichRow, 1:2],
                    collapse = ">")
return(edge)
}
```

In addition to the core sna and ggplot2 packages, the Hmisc package is used which provides the bezier() function used by edgeMaker.

```
library(sna)
library(ggplot2)
library(Hmisc)
```

As has been typical with the examples in this chapter, the network data has to be transformed to an edgelist format prior to using the plotting functions. For this example, we also drop the ties that have the associated weight 'passes' less than 10. Finally, the edgeMaker function expects the edgelist object to be named 'adjacencyList.'

```
data(FIFA_Nether)
fifa <- FIFA_Nether
fifa.edge <- as.edgelist.sna(fifa,attrname='passes')
fifa.edge <- data.frame(fifa.edge)
names(fifa.edge)[3] <- "value
fifa.edge <- fifa.edge[fifa.edge$value > 9,]
adjacencyList <- fifa.edge
```

Now, we use edgeMaker to create the curved edges. Also, gplot (from sna) is called once to store the layout coordinates for the ggplot2 function. (This means that any set of coordinates can be fed to ggplot2.)

```
layoutCoordinates <- gplot(network(fifa.edge))
allEdges <- lapply(1:nrow(fifa.edge),
                   edgeMaker, len = 500, curved = TRUE)
allEdges <- do.call(rbind, allEdges)
```

Before producing the plot, we create an empty ggplot2 theme. This is used to clean up after producing the plot.

```
new_theme_empty <- theme_bw()
new_theme_empty$line <- element_blank()
new_theme_empty$rect <- element_blank()
new_theme_empty$strip.text <- element_blank()
new_theme_empty$axis.text <- element_blank()
new_theme_empty$plot.title <- element_blank()
new_theme_empty$axis.title <- element_blank()
new_theme_empty$plot.margin <- structure(c(0,0,-1,-1),
```

```
                               unit = "lines",
                               valid.unit = 3L,
                               class = "unit")
```

And now the final step is to create the plot using `ggplot()`. Familiarity with ggplot2 will help in understanding this code. The `scale_colour_gradient` option controls the intensity of the gradient, and the `scale_size` option controls the amount of the taper (Fig. 6.6).

```
zp1 <- ggplot(allEdges)
zp1 <- zp1 + geom_path(aes(x = x, y = y, group = Group,
                       colour=Sequence, size=-Sequence))
zp1 <- zp1 + geom_point(data =
                          data.frame(layoutCoordinates),
                        aes(x = x, y = y),
                        size = 4, pch = 21,
                        colour = "black", fill = "gray")
zp1 <- zp1 + scale_colour_gradient(low = gray(0),
                              high = gray(9/10),
                              guide = "none")
zp1 <- zp1 + scale_size(range = c(1/10, 1.5),
                        guide = "none")
zp1 <- zp1 + new_theme_empty
print(zp1)
```

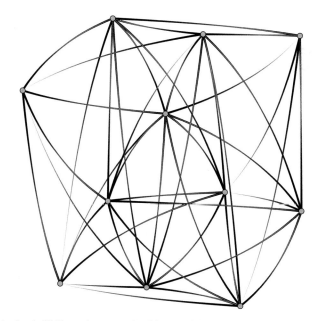

Fig. 6.6 Netherlands 2010 passing network with curved ties

Part III
Description and Analysis

Chapter 7
Actor Prominence

*...we're now tied up with human beings, tied to you and forced
to go on with this adventure according to the laws of visibility.*
(Jean Genet)

7.1 Introduction

Networks are interesting because of their specific structural patterns, and how those
structures affect the members of the network. Stated more simply, networks affect
their members based on where those members are located in the networks. A person
who is connected to many other members of a network is likely to view the rest
of the network quite differently from somebody who is relatively isolated from the
other members.

Network analysis provides many tools for viewing, analyzing, and assessing
the locations of individual nodes and ties. This is often the first type of network
analysis that is performed once network data are obtained, beyond simple network
description.

By examining the location of individual network members, we can assess the
prominence of those members. An actor is prominent if the ties of the actor make
that actor visible to the other members in the network (Knoke and Burt 1983). In
the rest of this chapter, we will cover a number of the most common ways to assess
network member prominence. For non-directed networks we will look at *centrality*;
where we view a central actor as one who is involved in many (direct or indirect) ties.
For directed networks, prominence is usually referred to as *prestige*; a prestigious
actor is one who is the object of extensive ties. This chapter will also cover how
individual node-level measures of centrality and prestige can be aggregated into
network-level *centralization* measures. An example of how to report the results of
prominence analysis will be presented. Finally, there will be a short discussion of
identifying cutpoints and bridges in networks. These are technically not measures
of prominence, but are simple locational properties of individual nodes or ties, and
as such are somewhat similar to the rest of the chapter's subject matter.

© Springer International Publishing Switzerland 2015
D.A. Luke, *A User's Guide to Network Analysis in R*, Use R!,
DOI 10.1007/978 3 319 23883 8_7

7.2 Centrality: Prominence for Undirected Networks

It makes intuitive sense that a network member who is connected to many other
members of the network is in a prominent position. For non-directed networks, we
will say that this type of actor has high centrality, or that it is in a central position.
However, there are a number of ways of operationalizing this type of prominence.
In fact, there are dozens of centrality statistics available to the network analyst.

To see how we can come up with different types of centrality measures, consider
the example network displayed in Fig. 7.1, based on the following simple sociomatrix, called net_mat.

```
##    a b c d e f g h i j
## a  0 1 1 0 0 0 0 0 0 0
## b  1 0 1 0 0 0 0 0 0 0
## c  1 1 0 1 1 0 1 0 0 0
## d  0 0 1 0 1 0 0 0 0 0
## e  0 0 1 1 0 1 0 0 0 0
## f  0 0 0 0 1 0 1 0 0 0
## g  0 0 1 0 0 1 0 1 0 0
## h  0 0 0 0 0 0 1 0 1 1
## i  0 0 0 0 0 0 0 1 0 0
## j  0 0 0 0 0 0 0 1 0 0
```

Which node is most central? Nodes c and g are both positioned in the center of
the graph, but as we learned in Chap. 4, the location of nodes in network graphics
may or may not hold any particular meaning. However, node c is directly connected
to more network members than any other node, so in that sense we could view c as
a central node. Alternatively, node g does not have as many direct network ties, but
it is positioned in such a way that it connects two different parts of the network. In
particular, the only way that information from nodes h, i, and j gets to the rest of the
network is through node g. Finally, even though node g is only directly connected
to two other nodes, it is positioned so that it is fairly close to every other node in the
network. Specifically, node g can reach every other node in only one or two steps.
That is, node g is connected to the rest of the network by paths of length one or two.
So, in these two very different senses, node g can also be thought of as a central
node.

In the next three sections, we will cover the three most commonly used measures
of centrality.

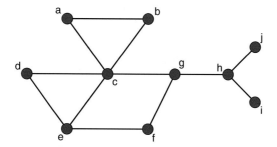

Fig. 7.1 Network graph example to demonstrate concepts of prominence

7.2.1 Three Common Measures of Centrality

7.2.1.1 Degree Centrality

The simplest measure of centrality by far is based on the notion that a node that has more direct ties is more prominent than nodes with fewer or no ties. Degree centrality thus, is simply the degree of each node. We first introduced node degree in Chap. 2. The degree of a node is the number of ties it has with other nodes.

Following the notation of Wasserman and Faust (1994), degree centrality is defined as:

$$C_D(n_i) = d(n_i)$$

The network in Fig. 7.1 is simple enough that we could count up the node degrees by hand. However, here is how degree centrality can be calculated in `statnet`, assuming that we have the data stored in a network object called `net`.

```
net <- network(net_mat)
net %v% 'vertex.names'

   ##   [1] "a" "b" "c" "d" "e" "f" "g" "h" "i" "j"

degree(net, gmode="graph")

   ##   [1] 2 2 5 2 3 2 3 3 1 1
```

The first line of code simply reminds you of the names of the nodes and their order. The `degree()` function calculates and returns the degree centrality scores for each node. The `gmode` option tells the function to treat the network object as a non-directed network (graph). (This option needs to be used, even if the network is created and stored as a non-directed network.)

The results confirm what we had already suggested above. Node *c* has the highest degree centrality. It is connected to five other nodes in the network, more than any other node.

7.2.1.2 Closeness Centrality

Instead of examining only the direct connections of the nodes, we can focus on how close each node is to every other node in a network. This leads to the concept of closeness centrality, where nodes are more prominent to the extent they are close to all other nodes in the network. Here is the relatively more complicated equation for closeness centrality,

$$C_C(n_i) = \left[\sum_{j=1}^{g} d(n_i, n_j) \right]^{-1}$$

where d is the path distance between two nodes. Closeness centrality, then, is the inverse of the sum of all the distances between node i and all the other nodes in the network.

```
closeness(net, gmode="graph")

##   [1] 0.409 0.409 0.600 0.429 0.450 0.450 0.600
##   [8] 0.474 0.333 0.333
```

This tells us that nodes c and g are tied with the highest closeness.

7.2.1.3 Betweenness Centrality

Betweenness centrality measures the extent that a node sits 'between' pairs of other nodes in the network, such that a path between the other nodes has to go through that node. A node with high betweenness is prominent, then, because that node is in a position to observe or control the flow of information in the network. The equation for betweenness centrality is

$$C_B(n_i) = \sum_{j<k} g_{jk}(n_i)/g_{jk}$$

where g_{jk} is the geodesic between nodes j and k. (A geodesic is the shortest path between two nodes.) $g_{jk}(n_i)$ is the number of geodesics between nodes j and k that contain node i.

```
betweenness(net, gmode="graph")

##   [1]   0.0   0.0  20.0   0.0   2.5   2.0  19.5  15.0   0.0
##  [10]   0.0
```

This shows that node c has the highest betweenness score, with nodes g and h not far behind. These quick examples show that different measures of centrality will emphasize different aspects of the prominence of nodes in a network.

7.2.2 Centrality Measures in R

R can handle many different measures of centrality and prestige. See the accompanying table for a list of the measures currently included in the statnet and igraph packages (Table 7.1).

As we can see, R provides a wide variety of ways to examine the centrality and prestige of individual actors in a network. The choice of which measure of centrality or prestige to use is driven in part by the type of network data you have; in particular, whether the network is directed or not. However as suggested in Sect. 7.2.1, the choice should be primarily driven by what type of information is provided by the particular prominence measure.

That being said, it is also useful to keep in mind that in many real-world social networks there is a great deal of overlap in the various centrality and prestige measures. Nodes that are identified as highly central using eigenvector centrality are also likely to be identified as central with other measures, especially those most closely related to eigenvector centrality (e.g., Bonacich power). We can illustrate this by showing the correlations among a set of centrality measures available in statnet applied to the DHHS Collaboration network.

Measures	statnet	igraph
Degree	degree()	degree()
Closeness	closeness()	closeness()
Betweenness	betweenness()	betweenness()
Eigenvector	evcent()	evcent()
Bonacich power	bonpow()	bonpow()
Flow betweenness	flowbet()	
Load	loadcent()	
Information	infocent()	
Stress	stresscent()	
Harary graph	graphcent()	
Bonacich alpha		alpha.centrality()
Kleinberg authority		auth.score()
Kleinberg hub		hub.score()
PageRank		page.rank()

Table 7.1 Prominence measures available in statnet and igraph

```
data(DHHS)
df.prom <- data.frame(
  deg = degree(DHHS),
  cls = closeness(DHHS),
  btw = betweenness(DHHS),
  evc = evcent(DHHS),
  inf = infocent(DHHS),
  flb = flowbet(DHHS)
)
```

```
cor(df.prom)
```

```
##         deg    cls    btw    evc    inf    flb
## deg  1.000  0.973  0.750  0.972  0.902  0.944
## cls  0.973  1.000  0.787  0.934  0.890  0.941
## btw  0.750  0.787  1.000  0.600  0.485  0.884
## evc  0.972  0.934  0.600  1.000  0.940  0.843
## inf  0.902  0.890  0.485  0.940  1.000  0.773
## flb  0.944  0.941  0.884  0.843  0.773  1.000
```

7.2.3 Centralization: Network Level Indices of Centrality

Centrality and prestige are characteristics of nodes in a network, based on the position of the node in the overall network. The variability of the individual centrality scores in a network can be very informative. For example, consider the following two extreme examples: a star graph, and a circle graph (Fig. 7.2).

```
dum1 <- rbind(c(1,2),c(1,3),c(1,4),c(1,5))
star_net <- network(dum1,directed=FALSE)
dum2 <- rbind(c(1,2),c(2,3),c(3,4),c(4,5),c(5,1))
circle_net <- network(dum2,directed=FALSE)
par(mar=c(4,4,.1,.1))
my_pal <- brewer.pal(5,"Set2")
gplot(star_net,usearrows=FALSE,displaylabels=FALSE,
      vertex.cex=2,
      vertex.col=my_pal[1],
      edge.lwd=0,edge.col="grey50",xlab="Star Graph")
gplot(circle_net,usearrows=FALSE,displaylabels=FALSE,
      vertex.cex=2,
      vertex.col=my_pal[3],
      edge.lwd=0,edge.col="grey50",xlab="Circle Graph")
```

In statnet, centralization is calculated using the centralization() function. The function accepts a name of an existing centrality or prestige function, and returns the appropriate network-level centralization score. Note that despite its name and the information presented in the help file for the function, centralization() can be used for directed graphs.

Using the star and circle graphs, we can see that every node has the same centrality score for the circle graph, leading to a minimum centralization score. The star graph shows the opposite pattern, where there is high variability between the node-level centrality scores, leading to higher centralization scores.

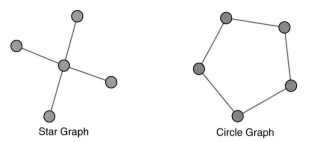

Star Graph   Circle Graph

Fig. 7.2 Centralization extreme examples

```
closeness(circle_net)

  ## [1] 0.667 0.667 0.667 0.667 0.667

centralization(circle_net,closeness)

  ## [1] 0

closeness(star_net)

  ## [1] 1.000 0.571 0.571 0.571 0.571

centralization(star_net,closeness)

  ## [1] 0.536
```

7.2.4 Reporting Centrality

All centrality and prestige functions in `statnet` (as well as `igraph`) produce a
vector of node-level scores, one for each actor in the network. Using the Bali terrorist network, we can see that centrality varies widely across the network members.

```
data(Bali)
str(degree(Bali))

  ## num [1:17] 18 8 18 30 18 20 6 18 18 10 ...

summary(degree(Bali))

  ##    Min. 1st Qu.  Median    Mean 3rd Qu.    Max.
  ##     6.0    10.0    18.0    14.8    18.0    30.0
```

These scores can be examined individually, but for both analysis and reporting, it
is usually more informative to examine patterns of prominence across nodes, across
different prominence measures, and even across different networks. In this section

we take a more in-depth look at the centrality of the actors in the Bali terrorist
network (Fig. 7.3).

```
data(Bali)
my_pal <- brewer.pal(5,"Set2")
rolecat <- Bali %v% "role"
gplot(Bali,usearrows=FALSE,displaylabels=TRUE,
      vertex.col=my_pal[as.factor(rolecat)],
      edge.lwd=0,edge.col="grey25")
legend("topright",legend=c("BM","CT","OA","SB",
                  "TL"),col=my_pal,pch=19,pt.cex=2)
```

If the network is small enough, it can be useful to examine the individual node-
level prominence scores. Table 7.2 shows the individual scores for each of three
common centrality measures: degree, closeness, and betweenness.

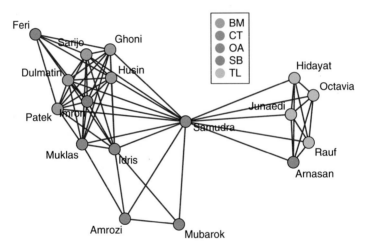

Fig. 7.3 Interactions among the 17 members of the 2002 Bali terrorist network

```
data(Bali)
df.prom2 <- data.frame(
  degree = degree(Bali),
  closeness = closeness(Bali),
  betweenness = betweenness(Bali)
  )
row.names(df.prom2) <- Bali %v% "vertex.names"
df.promsort <- df.prom2[order(-df.prom2$degree),]
cd <- centralization(Bali,degree)
cc <- centralization(Bali,closeness)
cb <- centralization(Bali,betweenness)
```

```
df.promsort <- rbind(df.promsort,c(cd,cc,cb))
row.names(df.promsort)[18] <- "\\emph{Centralization}"
```

	Degree	Closeness	Betweenness
Samudra	30.00	0.94	122.33
Idris	20.00	0.73	12.33
Muklas	18.00	0.70	4.67
Imron	18.00	0.70	3.33
Dulmatin	18.00	0.70	3.33
Husin	18.00	0.70	3.33
Ghoni	18.00	0.70	3.33
Patek	18.00	0.70	3.33
Sarijo	18.00	0.70	3.33
Feri	12.00	0.48	0.00
Arnasan	10.00	0.57	0.00
Rauf	10.00	0.57	0.00
Octavia	10.00	0.57	0.00
Hidayat	10.00	0.57	0.00
Junaedi	10.00	0.57	0.00
Amrozi	8.00	0.55	0.67
Mubarok	6.00	0.53	0.00
Centralization	0.54	0.33	0.50

Table 7.2 Centrality of the 17 members of the 2002 Bali terrorist network

As described in Chap. 5, nodes can be sized according to any informative quantitative characteristic of the actors. This can be non-network information such as age or weight. More useful here is to use information from the network itself; in this case the centrality scores for each node.

This can easily be done using the network plotting options. In fact, only one additional parameter (vertex.cex) needs to be passed to gplot(). This parameter can be a constant, in which case it simply controls the overall size of each vertex in the graph. However, you can also pass it a vector of numeric scores. All of the node-level prominence measures return a numeric vector, so that is what can be used to scale node size based on centrality or prestige.

The only tricky issue is that R reads the raw numbers passed to vertex.cex, and these numbers are often too small or too large. Typically, you will need to play around with some type of scaling factor to ensure that the graphic is interpretable. (See Chap. 5 for a more detailed treatment of node sizing and scaling.) The following two figures illustrate this. First, the normalized degree centrality for the Bali network is calculated (option rescale=TRUE). This rescales the raw degree scores so that they all fall between 0 and 1. The first figure shows that the normalized degree scores are too small. The second graph uses the same information, but the vertex.cex parameter is multiplied by 20 so that the relative differences between the node sizes can be seen (Fig. 7.4).

```
deg <- degree(Bali,rescale=TRUE)
op <- par(mfrow=c(1,2))
gplot(Bali,usearrows=FALSE,displaylabels=FALSE,
      vertex.cex=deg,
      vertex.col=my_pal[as.factor(rolecat)],
      edge.lwd=0,edge.col="grey25",
      main="Too small")
gplot(Bali,usearrows=FALSE,displaylabels=FALSE,
      vertex.cex=deg*20,
      vertex.col=my_pal[as.factor(rolecat)],
      edge.lwd=0,edge.col="grey25",
      main="A little better")
par(op)
```

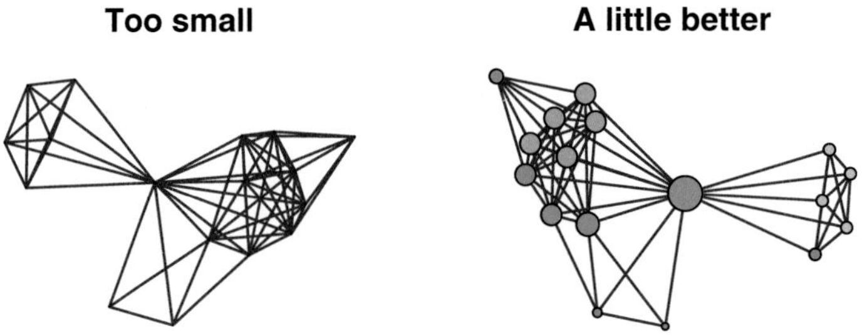

Fig. 7.4 Comparison of two approaches to sizing vertices by degree centrality

A network graphic that includes node-level prominence information can be an effective analysis and communication tool. The overall structure of the network can be made clear, as well as the importance of individual positions. Figure 7.5 is a final version of the Bali network graphic that combines node-level categorical information (denoted by vertex color) with node-level quantitative information (denoted by vertex size).

```
deg <- degree(Bali,rescale=TRUE)
gplot(Bali,usearrows=FALSE,displaylabels=TRUE,
      vertex.cex=deg*12,
      vertex.col=my_pal[as.factor(rolecat)],
      edge.lwd=0.5,edge.col="grey75")
legend("topright",legend=c("BM","CT","OA","SB","TL"),
       col=my_pal,pch=19,pt.cex=2)
```

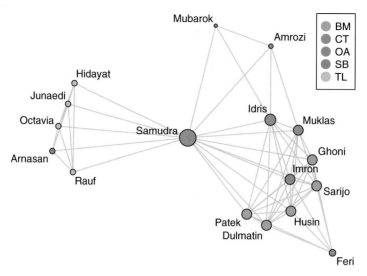

Fig. 7.5 Bali network with nodes sized according to degree centrality

7.3 Cutpoints and Bridges

There are two additional concepts from graph theory that can be useful tools when assessing locational properties of individual nodes or ties. The first is a *cutpoint*, which is defined as a node that, if dropped, would increase the number of components in the network. In many types of networks cutpoints thus occupy important positions connecting different parts of the network. If they were dropped, that would result in two subsets of actors that would not be able to communicate with each other (Fig. 7.6).

You can use the cutpoint() function in statnet to quickly identify any cutpoints in a network. (For a directed network, you need to specify whether you are using a 'weak' or 'strong' component rule for identifying cutpoints. See ?cutpoints for further information.)

```
cpnet <- cutpoints(net,mode="graph",
                   return.indicator=TRUE)
gplot(net,gmode="graph",vertex.col=cpnet+2,coord=coords,
      jitter=FALSE,displaylabels=TRUE)
```

So, in addition to the two central nodes (*c* and *g*) we had identified earlier, we can see that *h* is also a cutpoint. Although simple to see in this example, we can confirm the nodes as cutpoints in a few different ways (Fig. 7.7).

```
net2 <- net
components(net2)
   ## [1] 1
```

```
delete.vertices(net2,7)
components(net2)
```

```
## [1] 2
```

```
gplot(net2,gmode="graph",vertex.col=2,
      coord=coords[-7,],jitter=FALSE,displaylabels=TRUE)
```

Bridges are the edge equivalent to cutpoints. That is, an edge is a bridge if removing it will split one component into two. There is no bridge identification function
built into statnet, but it is relatively easy to create a function that will detect
bridges. This function takes a statnet directed or non-directed network, and ex-
amines each tie to see if removing it changes the component count. A logical vector
with length equal to the number of ties is returned indicating which ties are bridges.

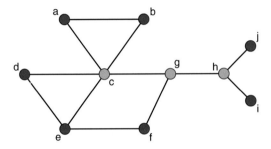

Fig. 7.6 Example graph with identified cutpoints

```
bridges <- function(dat,mode="graph",
                    connected=c("strong", "weak")) {
  e_cnt <- network.edgecount(dat)
  if (mode == "graph") {
    cmp_cnt <- components(dat)
    b_vec <- rep(FALSE,e_cnt)
    for(i in 1:e_cnt){
      dat2 <- dat
      delete.edges(dat2,i)
      b_vec[i] <- (components(dat2) != cmp_cnt)
      }
  }
  else {
    cmp_cnt <- components(dat,connected=connected)
    b_vec <- rep(FALSE,e_cnt)
    for(i in 1:e_cnt){
      dat2 <- dat
```

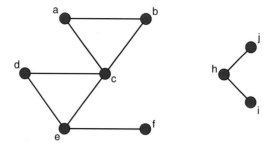

Fig. 7.7 Example graph with one cutpoint dropped

```
    delete.edges(dat2,i)
    b_vec[i] <- (components(dat2,connected=connected)
      != cmp_cnt)
  }
}
return(b_vec)
}
```

Once the function has been defined, we can use it directly:

```
bridges(net)
```

```
##  [1] FALSE FALSE FALSE FALSE FALSE FALSE FALSE
##  [8] FALSE FALSE FALSE FALSE FALSE FALSE FALSE
## [15] FALSE FALSE FALSE FALSE  TRUE  TRUE  TRUE
## [22]  TRUE  TRUE  TRUE
```

This shows us that there are three ties that are bridges in the example network.
We can also use the bridges function similarly to the cutpoints function in a graphic
to display which edges are bridges (Fig. 7.8).

```
brnet <- bridges(net)
gplot(net,gmode="graph",vertex.col="red",
      edge.col=brnet+2,coord=coords,
      jitter=FALSE,displaylabels=TRUE)
```

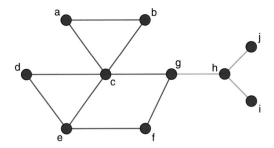

Fig. 7.8 Example graph with identified bridges

Chapter 8
Subgroups

> *Our young people are faced by a series of different groups*
> *which believe different things and advocate different practices,*
> *and to each of which some trusted friend or relative may belong.*
> (Margaret Mead)

8.1 Introduction

The social systems contained in networks often exhibit complex structures. For example, in his classic *The strength of weak ties*, Granovetter (1973) suggested that many social networks are made up of relatively densely connected subgroups (e.g., friendship subnetworks) that are themselves only connected via less common ties (e.g., between acquaintances). It then follows that it will be important to be able to define and identify such subgroups. Many disciplines have theories that assume that larger social systems are made up of distinguishable subgroups, for example sociologists consider social classes; psychologists examine small group behavior, and public health examine health disparities between different social groups.

This chapter covers a number of techniques available within R to identify and examine subgroups that may be contained in larger social networks. The `igraph` package is used extensively in this chapter, because of the depth of its coverage of subgroup and community detection techniques.

At times, it may not be necessary to use specific subgroup techniques. For example, reconsider Moreno's sociogram that we examined in Chap. 2 (Fig. 8.1). Here, it is self-evident that the network is made up of two primary groups, even if we did not know beforehand that this depicts a primary school class.

However, in most real-world social networks the subgroup structure is not as clear, if it even exists at all. Figure 8.2 shows a more realistic network, where the there is a hint of some subgroup structure, but more systematic analysis will be required to reveal it. The color coding and labels suggest that the there may be some cohesion among members from the same DHHS agency, but it is not crystal clear.

© Springer International Publishing Switzerland 2015   105
D.A. Luke, *A User's Guide to Network Analysis in R*, Use R!,
DOI 10.1007/978-3-319-23883-8_8

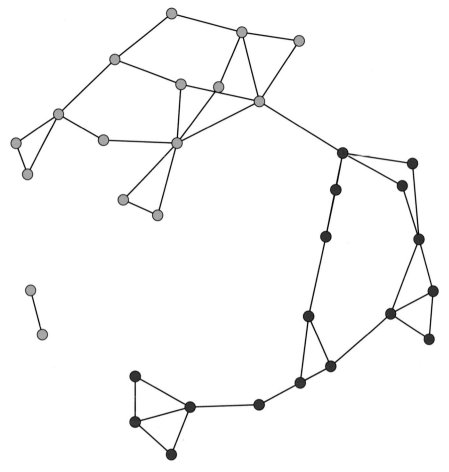

Fig. 8.1 Moreno sociogram, showing two subgroups

8.2 Social Cohesion

One way to think about network subgroups is through social cohesion. Cohesive subgroups are sets of actors that are tied together through frequent, strong, and direct ties (Wasserman and Faust 1994). This approach is so intuitive that it led to a number of the earliest techniques for identifying network subgroups.

8.2.1 Cliques

Cliques are one of the simplest types of cohesive subgroups, and because of their straightforward definition are also one of the easiest types to understand. A clique is a maximally complete subgraph; that is, it is a subset of nodes that have all possible ties among them.

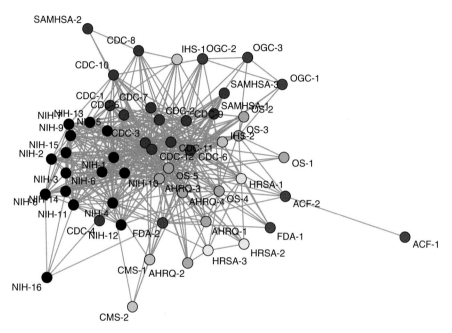

Fig. 8.2 Collaboration ties among DHHS agencies

Consider the example graph in Fig. 8.3. There are two cliques in this graph: A,B,C,D and E,F,G. (Technically, connected dyads also are cliques, but typically only cliques of size 3 or larger are of interest. Also, by definition any clique of size *k* will also contain all the cliques sized *k-1*, *k-2*, etc.)

```
library(igraph)
clqexmp <- graph.formula(A:B:C:D--A:B:C:D,D-E,E-F-G-E)
```

The following commands demonstrate how to get information about any cliques in a network. Despite what the name suggests, `clique.number()` does not return the number of cliques, but the size of the largest clique. To get a list of all the cliques, constrained by a minimum or maximum size, use `cliques()`. When there are a large number of cliques in a network, `maximal.cliques()` may be more useful. Finally, as the name suggests, `largest.cliques()` will find all of the largest cliques in a network.

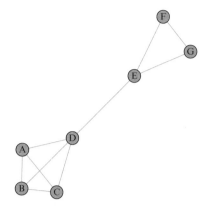

Fig. 8.3 Example graph with two cliques

```
clique.number(clqexmp)

  ## [1] 4

cliques(clqexmp, min=3)

  ## [[1]]
  ## + 3/7 vertices, named:
  ## [1] A B C
  ##
  ## [[2]]
  ## + 3/7 vertices, named:
  ## [1] A B D
  ##
  ## [[3]]
  ## + 3/7 vertices, named:
  ## [1] A C D
  ##
  ## [[4]]
  ## + 3/7 vertices, named:
  ## [1] B C D
  ##
  ## [[5]]
  ## + 3/7 vertices, named:
  ## [1] E F G
  ##
  ## [[6]]
  ## + 4/7 vertices, named:
  ## [1] A B C D

maximal.cliques(clqexmp,min=3)
```

```
## [[1]]
## + 3/7 vertices, named:
## [1] E F G
##
## [[2]]
## + 4/7 vertices, named:
## [1] A B D C
```

`largest.cliques(clqexmp)`

```
## [[1]]
## + 4/7 vertices, named:
## [1] D A B C
```

Note that the latter three functions return lists of vertex ids. When the `igraph` object has vertex names, the following syntax shows how names rather than ids can be displayed.

`V(clqexmp)[unlist(largest.cliques(clqexmp))]`

```
## + 4/7 vertices, named:
## [1] D A B C
```

Cliques, however, have two major disadvantages that reduce their utility in real-world social network analysis. First, a clique is a very conservative definition of a cohesive subgroup. Consider a subgraph made up of seven vertices. To be a clique, all of the 21 possible ties must exist between all seven members. If only one is missing, then the seven vertices will not belong to one clique, even though the density of these seven vertices ($20/21 = 0.95$) would suggest that this is a cohesive subgroup.

A consequence of this fragility is the second major issue of cliques: they simply are not very common in larger social networks. Table 8.1 presents some simple simulation results demonstrating the rarity of cliques. Four random networks were created with 25, 50, 100, and 500 nodes. For each network, the average degree was constrained to approximately 6. The table shows that the number of cliques remains roughly constant, even as the network size increases dramatically. Furthermore, the cliques remain small in size. (See Chap. 10 for more information on random graph models such as `erdos.renyi.game`.)

```
g25 <- erdos.renyi.game(25, 75, type="gnm")
g50 <- erdos.renyi.game(50, 150, type="gnm")
g100 <- erdos.renyi.game(100, 300, type="gnm")
g500 <- erdos.renyi.game(500, 1500, type="gnm")
nodes <- c(25,50,100,500)
lrgclq <- c(clique.number(g25),clique.number(g50),
           clique.number(g100),clique.number(g500))
numclq <- c(length(cliques(g25,min=3)),
           length(cliques(g50,min=3)),
```

```
          length(cliques(g100,min=3)),
          length(cliques(g500,min=3)))
clqinfo <- data.frame(Nodes=nodes,Largest=lrgclq,
                      Number=numclq)
```

Nodes	Largest	Number
25	3	27
50	4	47
100	3	35
500	3	38

Table 8.1 Demonstration of clique characteristics

8.2.2 k-Cores

Partly because of the rarity of cliques in observed social networks, a number of variations on the clique idea have been proposed. A popular variation is the *k-core*. A *k*-core is a maximal subgraph where each vertex is connected to at least *k* other vertices in the subgraph. *k*-cores have a number of advantages: they are nested (every member of a 4-core is also a member of a 3-core, and so on), they do not overlap, and they are easy to identify.

An analysis of *k*-cores typically proceeds by first identifying the entire *k*-core set, then doing visual examination of the *k*-core structures, possibly followed by closer examination of an individual *k*-core level (e.g., the 6-core).

For an example of how to examine *k*-cores in a social network we will use the DHHS dataset. This dataset shows collaboration ties among tobacco control experts working across various institutes and agencies within the Department of Health and Human Services in 2005 (Leischow, Luke, et al. 2010). This dataset is a statnet network object, so we start by translating it to an igraph object using the intergraph package. Also, DHHS has a valued tie, where the values range from 1 (only share information) to 4 (formal collaboration across multiple projects). As the density suggests, if all values are included then the network is highly interconnected. For this example we will only examine formal collaboration ties (i.e., the 'collab' edge attribute is 3 or 4). Here we use the igraph subgraph.edges function to select only those edges. This gives us a new network that is half as dense as the original.

```
data(DHHS)
library(intergraph)
iDHHS <- asIgraph(DHHS)
graph.density(iDHHS)
```

```
## [1] 0.312
```

```
iDHHS <- subgraph.edges(iDHHS,E(iDHHS)[collab > 2])
graph.density(iDHHS)
```

```
## [1] 0.153
```

To identify the *k*-core structure in the network, the `graph.coreness` function is used. It returns a vector listing the highest core that each vertex belongs to in the network. The results tell us the *k*-cores range from 1 to 6.

```
coreness <- graph.coreness(iDHHS)
table(coreness)
```

```
## coreness
## 1  2  3  4  5  6
## 7  6  2  5  2 26
```

```
maxCoreness <- max(coreness)
maxCoreness
```

```
## [1] 6
```

To better understand the *k*-core structure, we can plot the network using the *k*-core set information. This example illustrates how `igraph` uses the special vertex attributes `name` and `color`. The name attribute is used by default to label the nodes in a plot. Here we copy over the vertex names that were stored in the `statnet` vertex attribute `vertex.names`. The color vertex attribute is used to set the default colors of the nodes. Here we add 1 to the k-core values stored in the coreness vector as a quick and dirty way to pick different colors for each k-core, as well as to avoid black (Fig. 8.4).

```
Vname <- get.vertex.attribute(iDHHS,name='vertex.names',
                              index=V(iDHHS))
V(iDHHS)$name <- Vname
V(iDHHS)$color <- coreness + 1
op <- par(mar = rep(0, 4))
plot(iDHHS,vertex.label.cex=0.8)
par(op)
```

To help with the interpretation, we can label the nodes with their *k*-core membership value. Also, in this example we demonstrate an alternative way to automatically pick a distinctive set of colors for the nodes (Fig. 8.5).

```
colors <- rainbow(maxCoreness)
op <- par(mar = rep(0, 4))
plot(iDHHS,vertex.label=coreness,
     vertex.color=colors[coreness])
par(op)
```

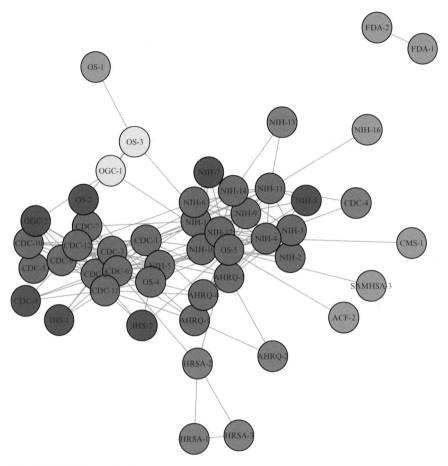

Fig. 8.4 DHHS k-core structure

This figure shows that the center of the network is made up primarily of the highest k-core. In this case, the 6-core is comprised of 26 of the 54 total nodes. Because of the nested structure of k-cores, we can further examine the subgroup patterns by progressively 'peeling away' each of the lower k-cores in turn. To do this we take advantage of the induced.subgraph function (Fig. 8.6).

```
V(iDHHS)$name <- coreness
V(iDHHS)$color <- colors[coreness]
iDHHS1_6 <- iDHHS
iDHHS2_6 <- induced.subgraph(iDHHS,
                             vids=which(coreness > 1))
iDHHS3_6 <- induced.subgraph(iDHHS,
                             vids=which(coreness > 2))
iDHHS4_6 <- induced.subgraph(iDHHS,
                             vids=which(coreness > 3))
```

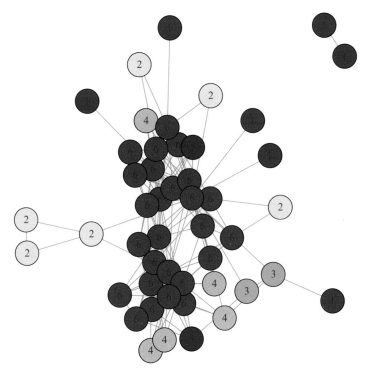

Fig. 8.5 DHHS k-core structure, with k-core membership values

```
iDHHS5_6 <- induced.subgraph(iDHHS,
                              vids=which(coreness > 4))
iDHHS6_6 <- induced.subgraph(iDHHS,
                              vids=which(coreness > 5))


lay <- layout.fruchterman.reingold(iDHHS)
op <- par(mfrow=c(3,2),mar = c(3,0,2,0))
plot(iDHHS1_6,layout=lay,main="All k-cores")
plot(iDHHS2_6,layout=lay[which(coreness > 1),],
     main="k-cores 2-6")
plot(iDHHS3_6,layout=lay[which(coreness > 2),],
     main="k-cores 3-6")
plot(iDHHS4_6,layout=lay[which(coreness > 3),],
     main="k-cores 4-6")
plot(iDHHS5_6,layout=lay[which(coreness > 4),],
     main="k-cores 5-6")
plot(iDHHS6_6,layout=lay[which(coreness > 5),],
     main="k-cores 6-6")
par(op)
```

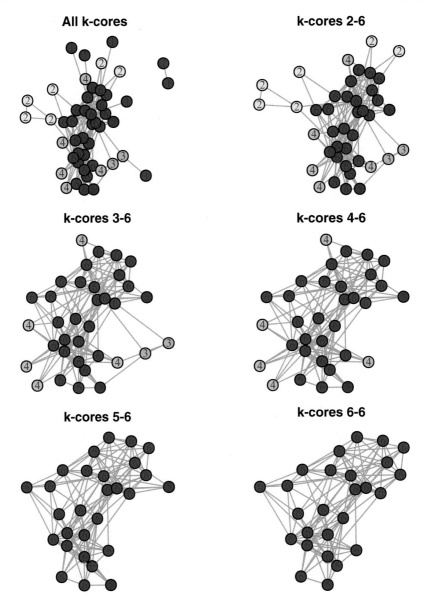

Fig. 8.6 Peeling away DHHS k-cores

8.3 Community Detection

Both cliques and k-cores are examples of subgroup identification that rely entirely on the pattern of internal ties defining the particular groups. Network scientists have developed a wide variety of subgroup identification algorithms and heuristics that define groups not just based on the internal ties, but also the pattern of ties between different groups. That is, a subgroup in a network is a set of nodes that has a relatively large number of internal ties, and also relatively few ties from the group to other parts of the network. These approaches vary in their details, but they are all designed to identify internally cohesive subgroups that are somewhat separated or isolated from other groups or nodes. These approaches are sometimes called community detection algorithms.

8.3.1 Modularity

An important characteristic of a network that is used in many community detection algorithms is that of *modularity*. Modularity is a measure of the structure of the network, specifically the extent to which nodes exhibit clustering where there is greater density within the clusters and less density between them (Newman 2006). Modularity can be used in an exploratory fashion, where an algorithm tries to maximize modularity and returns the node classification that is found to best explain the observed clustering. Conversely, modularity can be used in a descriptive fashion where the modularity statistic is calculated for any node classification variable of interest. For example, an analyst can calculate the modularity score for a friendship network given the gender of network members. Used this way, modularity reflects the extent to which gender explains the observed clustering among the friends in the network.

Modularity is a chance-corrected statistic, and is defined as the fraction of ties that fall within the given groups minus the expected such fraction if ties were distributed at random. The modularity statistic can range from $-1/2$ to $+1$. The closer to 1, the more the network exhibits clustering with respect to the given node grouping.

Consider the following simple example of a network with nine nodes. We have two categorical vertex attributes which each classify the nodes into three groups (Fig. 8.7).

```
g1 <- graph.formula(A-B-C-A,D-E-F-D,G-H-I-G,A-D-G-A)
V(g1)$grp_good <- c(1,1,1,2,2,2,3,3,3)
V(g1)$grp_bad <- c(1,2,3,2,3,1,3,1,2)


op <- par(mfrow=c(1,2))
plot(g1,vertex.color=(V(g1)$grp_good),
     vertex.size=20,
```

```
      main="Good Grouping")
plot(g1,vertex.color=(V(g1)$grp_bad),
      vertex.size=20,
      main="Bad Grouping")
par(op)
```

Good Grouping   **Bad Grouping**

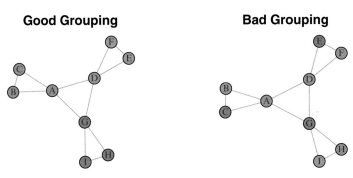

Fig. 8.7 Modularity example

As the figure suggests, the clustering that is evident in the network is better
accounted for by the grp_good node attribute compared to the grp_bad variable. This can be confirmed by calculating the modularity score provided by the
modularity function in igraph.

```
modularity(g1,V(g1)$grp_good)
```

```
  ## [1]  0.417
```

```
modularity(g1,V(g1)$grp_bad)
```

```
  ## [1]  -0.333
```

Real-world social networks are often characterized by clustering, but it is of
course harder to judge the extent of the clustering by eye. Earlier in the chapter
we saw that there was interesting subgroup structure contained in the DHHS network, and that this structure might be partially explained by the DHHS organization
that the person worked for.

```
library(intergraph)
data(DHHS)
iDHHS <- asIgraph(DHHS)
table(V(iDHHS)$agency)
```

```
  ##
  ##  0  1  2  3  4  5  6  7  8  9 10
  ##  2  4 12  2  2  3  2 16  3  5  3
```

```
V(iDHHS)[1:10]$agency
```

```
## [1] 0 0 1 1 1 1 2 2 2 2
```

```
modularity(iDHHS,(V(iDHHS)$agency+1))
```

```
## [1] 0.14
```

The `modularity` function expects that the node grouping variable is numbered starting at 1 (and in fact the current version of the function will crash if community membership has zeros.) In this case, `agency` is numbered starting at 0 so we add 1 to it before passing it to the `modularity` function. The modularity score of 0.14 suggests that the DHHS agency does explain some of the clustering that is present in the network. However, like most network descriptive statistics, the number in and of itself has little meaning. The interpretation of the network characteristic becomes more meaningful when it is compared to another relevant measure. For example, how does the modularity change over time? Or, how does this modularity score compare to the modularity score for a different vertex attribute on the same network?

As a comparison, we can look at the modularity scores for two other datasets included in `UserNetR`. For the `Moreno` data, we can see how `gender` accounts for subgroup structure. For the `Facebook` network, we can use the `group` node attribute, which designates the type of social group the Facebook friends belong to (family, work, music, high school, etc.). The results show us that both the Moreno and Facebook social networks exhibit higher modularity than the DHHS network.

```
data(Moreno)
iMoreno <- asIgraph(Moreno)
table(V(iMoreno)$gender)
```

```
##
## 1 2
## 16 17
```

```
modularity(iMoreno,V(iMoreno)$gender)
```

```
## [1] 0.476
```

```
data(Facebook)
levels(factor(V(Facebook)$group))
```

```
## [1] "B" "C" "F" "G" "H" "M" "S" "W"
```

```
grp_num <- as.numeric(factor(V(Facebook)$group))
modularity(Facebook,grp_num)
```

```
## [1] 0.615
```

8.3.2 Community Detection Algorithms

The main reason that this chapter uses `igraph` is that it includes support for many if not most of the existing community detection approaches. Table 8.2 lists the currently supported algorithms, along with whether each function supports directed networks, weighted networks (networks with valued ties), and whether the algorithm can be used on networks with more than one component.

Name	Function	Directed	Weighted	Components
Edge-betweenness	cluster_edge_betweenness	T	T	T
Leading eigenvector	cluster_leading_eigen	F	F	T
Fast-greedy	cluster_fast_greedy	F	T	T
Louvain	cluster_louvain	F	T	T
Walktrap	cluster_walktrap	F	T	F
Label propagation	cluster_label_prop	F	T	F
InfoMAP	cluster_infomap	T	T	T
Spinglass	cluster_spinglass	F	T	F
Optimal	cluster_optimal	F	T	T

Table 8.2 Community detection functions in igraph

The basic workflow for conducting community detection in `igraph` is to run one of the community detection functions on a network and store the results in a `communities` class object. Then, the identified subgroups in the network can be explored using a number of `igraph` functions that know how to operate with `communities` objects. The networks can also be plotted easily to show the results of the community detection.

For example, consider the simple Moreno friendship network that is clearly divided into two subgroups based on gender. Community detection on this network proceeds as follows.

```
cw <- cluster_walktrap(iMoreno)
membership(cw)

   ##   [1]  1 1 1 1 1 1 1 1 3 3 3 5 5 5 5 1 3 2 2 2 4 4 4
   ##  [24]  2 2 2 2 2 2 2 2 6 6

modularity(cw)

   ##  [1]  0.618
```

Modularity is fairly high, suggesting that the walktrap algorithm has done a good job at detecting subgroup structure. The membership function reveals that six different subgroups have been identified. These are best understood through visualization. If the plot function is passed a communities object along with the network it belongs to, then an attractive plot is produced with each subgroup getting its own color-coded shaded polygon (Fig. 8.8).

```
plot(cw, iMoreno)
```

Once you have a community detection subgroup solution, it can be examined like any membership vector. For example, you can compare it to an existing partition based on a node characteristic. In this case, how does a walktrap solution compare to the specific agency of the nodes in the DHHS network?

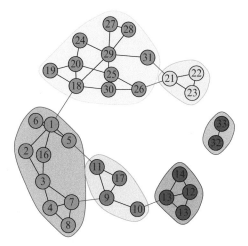

Fig. 8.8 Community detection on Moreno network

```
cw <- cluster_walktrap(iDHHS)
modularity(cw)

  ## [1] 0.165

membership(cw)

  ##  [1] 4 5 1 1 1 1 1 1 1 1 1 1 1 1 1 1 1 1 1 1 1 1 1
  ## [24] 1 1 1 1 3 3 3 3 3 3 3 3 3 3 3 3 3 3 3 3 3 2 2 2
  ## [47] 2 1 2 1 1 1 1 1

table(V(iDHHS)$agency,membership(cw))
```

	1	2	3	4	5
0	0	0	0	1	1
1	4	0	0	0	0
2	12	0	0	0	0
3	2	0	0	0	0
4	2	0	0	0	0
5	3	0	0	0	0

```
## 6     2    0    0    0    0
## 7     0    0   16    0    0
## 8     0    3    0    0    0
## 9     3    2    0    0    0
## 10    3    0    0    0    0
```

A common practice is to use more than one community detection algorithm and compare the results. (Remember that only some algorithms can handle particular types of networks such as directed networks.) The following examples explore how different algorithms find slightly different subgroups in the Bali terrorism network.

```
data(Bali)
iBali <- asIgraph(Bali)


cw <- cluster_walktrap(iBali)
modularity(cw)

  ## [1] 0.283

membership(cw)

  ## [1] 2 1 2 1 2 2 1 2 2 3 3 3 3 3 2 2 2

ceb <- cluster_edge_betweenness(iBali)
modularity(ceb)

  ## [1] 0.239

membership(ceb)

  ## [1] 1 1 1 1 1 1 1 1 1 2 2 2 2 2 1 1 1

cs <- cluster_spinglass(iBali)
modularity(cs)

  ## [1] 0.297

membership(cs)

  ## [1] 1 2 1 3 1 1 2 1 1 3 3 3 3 3 1 1 1

cfg <- cluster_fast_greedy(iBali)
modularity(cfg)

  ## [1] 0.263

membership(cfg)

  ## [1] 2 2 1 2 1 2 2 1 1 3 3 3 3 3 1 1 1
```

```
clp <- cluster_label_prop(iBali)
modularity(clp)
```

```
   ## [1] 0.239
```

```
membership(clp)
```

```
   ##  [1] 1 1 1 1 1 1 1 1 1 2 2 2 2 2 1 1 1
```

```
cle <- cluster_leading_eigen(iBali)
modularity(cle)
```

```
   ## [1] 0.275
```

```
membership(cle)
```

```
   ##  [1] 1 1 1 2 1 1 2 1 1 2 2 2 2 2 1 1 1
```

```
cl <- cluster_louvain(iBali)
modularity(cl)
```

```
   ## [1] 0.297
```

```
membership(cl)
```

```
   ##  [1] 3 1 3 2 3 3 1 3 3 2 2 2 2 2 3 3 3
```

```
co <- cluster_optimal(iBali)
modularity(co)
```

```
   ## [1] 0.297
```

```
membership(co)
```

```
   ##  [1] 1 2 1 3 1 1 2 1 1 3 3 3 3 3 1 1 1
```

These results show that all the detection algorithms identify either two or three subgroups. Modularity ranges from about 0.24 to 0.30.

The community detection results can be compared to one another using a number of classification comparison metrics, including the adjusted Rand statistic.

```
table(V(iBali)$role,membership(cw))
```

```
   ##
   ##       1 2 3
   ##    BM 0 5 0
   ##    CT 1 2 0
   ##    OA 2 1 0
   ##    SB 0 1 1
   ##    TL 0 0 4
```

```
compare(as.numeric(factor(V(iBali)$role)),cw,
        method="adjusted.rand")

  ## [1] 0.35

compare(cw,ceb,method="adjusted.rand")

  ## [1] 0.616

compare(cw,cs,method="adjusted.rand")

  ## [1] 0.89

compare(cw,cfg,method="adjusted.rand")

  ## [1] 0.669
```

Finally, we can plot multiple solutions to better understand the similarities and differences among the different community detection algorithms (Fig. 8.9).

```
op <- par(mfrow=c(3,2),mar=c(3,0,2,0))
plot(ceb, iBali,vertex.label=V(iBali)$role,
     main="Edge Betweenness")
plot(cfg, iBali,vertex.label=V(iBali)$role,
     main="Fastgreedy")
plot(clp, iBali,vertex.label=V(iBali)$role,
     main="Label Propagation")
plot(cle, iBali,vertex.label=V(iBali)$role,
     main="Leading Eigenvector")
plot(cs,  iBali,vertex.label=V(iBali)$role,
     main="Spinglass")
plot(cw,  iBali,vertex.label=V(iBali)$role,
     main="Walktrap")
par(op)
```

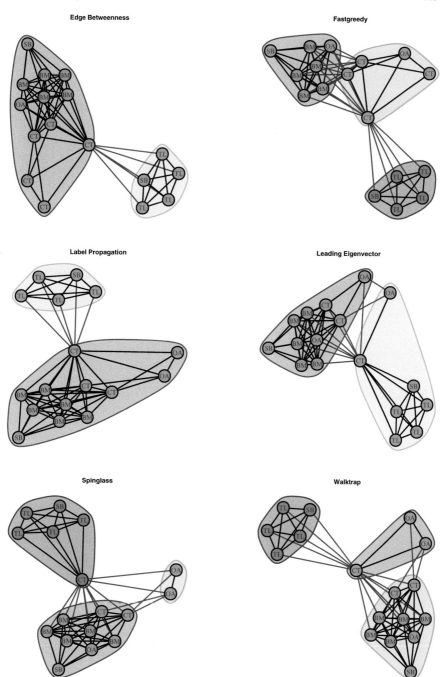

Fig. 8.9 Community detection comparisons on Bali network

Chapter 9
Affiliation Networks

> *A tribe is a group of people connected to one another, connected to a leader, and connected to an idea. For millions of years, human beings have been part of one tribe or another. A group needs only two things to be a tribe: a shared interest and a way to communicate.* (Seth Godin – *Tribes: We Need You to Lead Us*)

9.1 Defining Affiliation Networks

Until now, all the networks that we have examined are based on direct ties. That is, the social ties connecting the actors in the social network have been confirmed through self-report, direct observation, or some other type of data collection that tells us how actors are directly connected to one another.

However, social scientists are often interested in situations where there may be the opportunity for social relationships, but these relationships cannot be directly observed. However, by virtue of occupying the same social situation, we may infer that there is an opportunity or potential for social connections.

We call this new type of social network an *affiliation network*. An affiliation network is a network where the members are affiliated with one another based on co-membership in a group, or co-participation in some type of event. For example, students who all belong to the same class can be thought of as being connected to one another, although we may not know whether they actually have direct social ties.

The classic example from network science is the case of corporate interlocks. Company directors have the opportunity to interact with each other when they sit together on the same corporate board of directors. Moreover, the companies themselves can be seen to be connected through their shared director memberships. That is, when the same director sits on two different company boards, those companies are connected through that director. Sociologists and political scientists have used these types of affiliation networks to explain how companies tend to behave in similar ways to one another (Galaskiewicz 1985).

© Springer International Publishing Switzerland 2015
D.A. Luke, *A User's Guide to Network Analysis in R*, Use R!,
DOI 10.1007/978-3-319-23883-8_9

9.1.1 Affiliations as 2-Mode Networks

As a simple example of an affiliation network, consider the following data table of
students grouped in classes (Table 9.1).

```
C1 <- c(1,1,1,0,0,0)
C2 <- c(0,1,1,1,0,0)
C3 <- c(0,0,1,1,1,0)
C4 <- c(0,0,0,0,1,1)
aff.df <- data.frame(C1,C2,C3,C4)
row.names(aff.df) <- c("S1","S2","S3","S4","S5","S6")
```

	C1	C2	C3	C4
S1	1	0	0	0
S2	1	1	0	0
S3	1	1	1	0
S4	0	1	1	0
S5	0	0	1	1
S6	0	0	0	1

Table 9.1 Students grouped by classes

This type of data matrix is called an *incidence matrix*, and it depicts how n actors
belong to g groups. In this case we have six students grouped into four classes.
An incidence matrix is similar to an adjacency matrix, but an adjacency matrix is
an nxn square matrix where each dimension refers to the actors in the network. An
incidence matrix, on the other hand, is an nxg rectangular matrix with two different
dimensions: actors and groups. For this reason, affiliation networks are also known
as *two-mode networks*.

9.1.2 Bipartite Graphs

In affiliation networks, there are always two types of nodes: one type for the actors,
and another type for the groups or events to which the actors belong. Ties then
connect the actors to those groups. One consequence of this is that there are no direct
ties among actors, and there are no direct ties between the groups. Figure 9.1, which
shows the example student affiliation network, illustrates this defining characteristic
of a bipartite graph.

Both `statnet` and `igraph` have functionality built in to recognize and operate
on affiliation networks. The process with `igraph` is a little more straightforward,
so it will be used in this chapter.

```
library(igraph)
bn <- graph.incidence(aff.df)


plt.x <- c(rep(2,6),rep(4,4))
plt.y <- c(7:2,6:3)
lay <- as.matrix(cbind(plt.x,plt.y))


shapes <- c("circle","square")
colors <- c("blue","red")
plot(bn,vertex.color=colors[V(bn)$type+1],
     vertex.shape=shapes[V(bn)$type+1],
     vertex.size=10,vertex.label.degree=-pi/2,
     vertex.label.dist=1.2,vertex.label.cex=0.9,
     layout=lay)
```

9.2 Affiliation Network Basics

9.2.1 Creating Affiliation Networks from Incidence Matrices

An affiliation network can be stored as an igraph object in a few different ways.
If the underlying data are available as an incidence matrix (e.g., Table 9.1), then this
can be done in one line of code.

```
bn <- graph.incidence(aff.df)
bn

 ## IGRAPH UN-B 10 11 --
 ## + attr: type (v/l), name (v/c)
 ## + edges (vertex names):
 ##   [1] S1--C1 S2--C1 S2--C2 S3--C1 S3--C2 S3--C3
 ##   [7] S4--C2 S4--C3 S5--C3 S5--C4 S6--C4
```

The graph.incidence function takes a matrix or data.frame and transforms
it into an affiliation network, reading the rows as actors and the columns as the
groups or events. Note that both the row and column names should be defined so
that they can be correctly assigned to the individual actors and groups.

By typing in the name of the igraph object, we get a cryptic two line summary
of the network. The 'B' in the 'UN-B' string tells us that this is a bipartite network. Furthermore, the second line shows that this network has two vertex attributes:

name stores the name of the vertex, and type is a logical vector that igraph uses
to distinguish between the two different types of nodes (students and classes, in this
case).

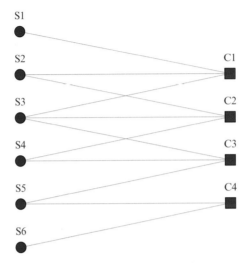

Fig. 9.1 Affiliation network as bipartite graph

More information about the affiliation network can be obtained using traditional
igraph functions.

```
get.incidence(bn)
```

```
##      C1 C2 C3 C4
## S1   1  0  0  0
## S2   1  1  0  0
## S3   1  1  1  0
## S4   0  1  1  0
## S5   0  0  1  1
## S6   0  0  0  1
```

```
V(bn)$type
```

```
##   [1]  FALSE FALSE FALSE FALSE FALSE FALSE    TRUE
##   [8]   TRUE   TRUE   TRUE
```

```
V(bn)$name
```

```
##   [1]  "S1" "S2" "S3" "S4" "S5" "S6" "C1" "C2" "C3"
## [10]  "C4"
```

Here we can see the underlying incidence matrix, and the vertex names and types. The correspondence between the names and the logical type vectors is clear, where all the students have type = FALSE, and the class nodes have type = TRUE.

9.2.2 Creating Affiliation Networks from Edge Lists

For larger networks, it is more common to have the underlying data available as an edge list. Edge lists can also be translated into an affiliation network, as long nodes of one type (e.g., students) are only connected to nodes of the other type (e.g., classes). The following code constructs the same example affiliation network from edge list data.

```
el.df <- data.frame(rbind(c("S1","C1"),
                  c("S2","C1"),
                  c("S2","C2"),
                  c("S3","C1"),
                  c("S3","C2"),
                  c("S3","C3"),
                  c("S4","C2"),
                  c("S4","C3"),
                  c("S5","C3"),
                  c("S5","C4"),
                  c("S6","C4")))
el.df

##       X1 X2
## 1    S1 C1
## 2    S2 C1
## 3    S2 C2
## 4    S3 C1
## 5    S3 C2
## 6    S3 C3
## 7    S4 C2
## 8    S4 C3
## 9    S5 C3
## 10   S5 C4
## 11   S6 C4

bn2 <- graph.data.frame(el.df,directed=FALSE)
bn2

## IGRAPH UN-- 10 11 --
## + attr: name (v/c)
## + edges (vertex names):
```

```
##   [1] S1--C1 S2--C1 S2--C2 S3--C1 S3--C2 S3--C3
##   [7] S4--C2 S4--C3 S5--C3 S5--C4 S6--C4
```

The above creates an network object, but `igraph` does not know that it is a bipartite graph. (Note that the network description lacks the 'B' that indicates a bipartite graph.) To fix this, we can simply set the `type` vertex attribute. Once this is done, we have an affiliation network object that is formed from a bipartite graph.

```
V(bn2)$type <- V(bn2)$name %in% el.df[,1]
bn2
```

```
## IGRAPH UN-B 10 11 --
## + attr: name (v/c), type (v/l)
## + edges (vertex names):
##   [1] S1--C1 S2--C1 S2--C2 S3--C1 S3--C2 S3--C3
##   [7] S4--C2 S4--C3 S5--C3 S5--C4 S6--C4
```

```
graph.density(bn)==graph.density(bn2)
```

```
## [1] TRUE
```

9.2.3 Plotting Affiliation Networks

As with any type of network, affiliation networks can be plotted for visual inspection. However, it is useful to designate different node shapes and colors to make the affiliation structure easier to interpret. Here we will set the students to be blue circles, and the classes to be red squares. The code shows how the `type` attribute can be used as an index into both a shapes and a colors vector to select the appropriate shape and color for each node. Note that 1 is added to `type` index because as a logical vector it starts at 0, whereas we want to select either the first or second elements of the shapes/colors vectors.

```
shapes <- c("circle","square")
colors <- c("blue","red")
plot(bn,vertex.color=colors[V(bn)$type+1],
     vertex.shape=shapes[V(bn)$type+1],
     vertex.size=10,vertex.label.degree=-pi/2,
     vertex.label.dist=1.2,vertex.label.cex=0.9)
```

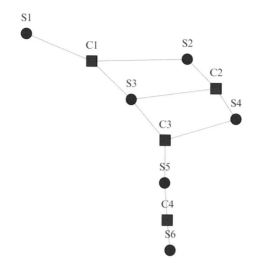

Fig. 9.2 Simple plot of affiliation network

9.2.4 Projections

By examining Fig. 9.2 we can see how classes are indirectly connected through their shared students. For example, classes 2 and 3 are indirectly connected through two shared students (S3 and S4). In comparison, classes 3 and 4 are also indirectly connected, but only via one shared student (S5). We can also focus the examination on the individual-level nodes. Here the figure reveals that students 2 and 4 are indirectly connected by their co-affiliation in class 2.

Examining both types of nodes in a two-mode network graphic is often the first step in studying an affiliation network. However, it is also useful to examine the direct connections among the nodes of one type at a time (classes and students, in this case). This can be done by extracting and visualizing the one-mode projections of the two-mode affiliation network. Every actor by event affiliation network can produce two one-mode networks, one of actors and one of the events or affiliations.

In igraph the projections can be obtained again by just one line of code. The bipartite.projection function returns a list of two igraph network objects. The first network is made up of the direct ties among the first mode (in our case students), and the second network shows the ties among the second mode (classes).

```
bn.pr <- bipartite.projection(bn)
bn.pr

  ## $proj1
  ## IGRAPH UNW- 6 8 --
  ## + attr: name (v/c), weight (e/n)
  ## + edges (vertex names):
```

```
## [1] S1--S2 S1--S3 S2--S3 S2--S4 S3--S4 S3--S5
## [7] S4--S5 S5--S6
##
## $proj2
## IGRAPH UNW- 4 4 --
## + attr: name (v/c), weight (e/n)
## + edges (vertex names):
## [1] C1--C2 C1--C3 C2--C3 C3--C4
```

Each of the list members can be accessed and treated like a typical igraph
network object, either within the list, or by extracting the list member.

```
graph.density(bn.pr$proj1)
```

```
## [1] 0.533
```

```
bn.student <- bn.pr$proj1
bn.class <- bn.pr$proj2
graph.density(bn.student)
```

```
## [1] 0.533
```

The adjacency matrix of each one-mode projection can be obtained with the
get.adjacency function. In the code below, notice how the edge attribute
weight is specified. This produces a valued adjacency matrix, where the values
indicate how many ties connect any of the nodes. So, for example, the Class adjacency matrix indicates that classes 2 and 3 have a weight of 2. This reflects the
observation we made earlier that classes 2 and 3 share two students (S3 and S4).

```
get.adjacency(bn.student,sparse=FALSE,attr="weight")
```

```
##    S1 S2 S3 S4 S5 S6
## S1  0  1  1  0  0  0
## S2  1  0  2  1  0  0
## S3  1  2  0  2  1  0
## S4  0  1  2  0  1  0
## S5  0  0  1  1  0  1
## S6  0  0  0  0  1  0
```

```
get.adjacency(bn.class,sparse=FALSE,attr="weight")
```

```
##    C1 C2 C3 C4
## C1  0  2  1  0
## C2  2  0  2  0
## C3  1  2  0  1
## C4  0  0  1  0
```

Each of the one-mode projections can, of course, be plotted for visual examination. Additionally, we can (at least for smaller networks) take advantage of the `weight` edge attribute to explore the relative strengths of the ties (Fig. 9.3).

```
shapes <- c("circle","square")
colors <- c("blue","red")
op <- par(mfrow=c(1,2))
plot(bn.student,vertex.color="blue",
     vertex.shape="circle",main="Students",
     edge.width=E(bn.student)$weight*2,
     vertex.size=15,vertex.label.degree=-pi/2,
     vertex.label.dist=1.2,vertex.label.cex=1)
plot(bn.class,vertex.color="red",
     vertex.shape="square",main="Classes",
     edge.width=E(bn.student)$weight*2,
     vertex.size=15,vertex.label.degree=-pi/2,
     vertex.label.dist=1.2,vertex.label.cex=1)
par(op)
```

9.3 Example: Hollywood Actors as an Affiliation Network

The data file hwd in the UseNetR package contains a larger and more interesting affiliation network that can be explored using these techniques. Hollywood actors are a good example of an affiliation network, actors are connected to one another through the movies in which they appear together. The hwd dataset is an igraph bipartite graph object. The data are originally from IMDB (www.imdb.com). The dataset contains the ten most popular movies (as judged by IMBD users) for

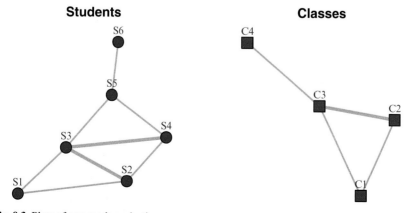

Fig. 9.3 Plots of one-mode projections

each year from 1999 to 2014, and the first ten actors listed on each movie's IMDB page. In addition to the movie and actor names, each movie has the year of its release, its IMDB user rating, and the MPAA movie rating (i.e., G, PG, PG-13, and R) stored as a node characteristic.

9.3.1 Analysis of Entire Hollywood Affiliation Network

The first steps to analyze these data are to load the file, and explore the basic affiliation structure of the network.

```
data(hwd)
h1 <- hwd
h1
```

```
## IGRAPH UN-B 1365 1600 --
## + attr: name (v/c), type (v/l), year (v/n),
## | IMDBrating (v/n), MPAArating (v/c)
## + edges (vertex names):
## [1] Inception--Leonardo DiCaprio
## [2] Inception--Joseph Gordon-Levitt
## [3] Inception--Ellen Page
## [4] Inception--Tom Hardy
## [5] Inception--Ken Watanabe
## [6] Inception--Dileep Rao
## [7] Inception--Cillian Murphy
## + ... omitted several edges
```

```
V(h1)$name[1:10]
```

```
## [1] "Inception"
## [2] "Alice in Wonderland"
## [3] "Kick-Ass"
## [4] "Toy Story 3"
## [5] "How to Train Your Dragon"
## [6] "Despicable Me"
## [7] "Scott Pilgrim vs. the World"
## [8] "Hot Tub Time Machine"
## [9] "Harry Potter and the Deathly Hallows: Part 1"
## [10] "Tangled"
```

```
V(h1)$type[1:10]
```

```
## [1] TRUE TRUE TRUE TRUE TRUE TRUE TRUE TRUE TRUE
## [10] TRUE
```

```
V(h1)$IMDBrating[1:10]
```

```
##  [1] 8.8 6.5 7.8 8.4 8.2 7.7 7.5 6.5 7.7 7.9
```

```
V(h1)$name[155:165]
```

```
##  [1] "Notting Hill"
##  [2] "Eyes Wide Shut"
##  [3] "The Green Mile"
##  [4] "10 Things I Hate About You"
##  [5] "American Pie"
##  [6] "Girl, Interrupted"
##  [7] "Leonardo DiCaprio"
##  [8] "Joseph Gordon-Levitt"
##  [9] "Ellen Page"
## [10] "Tom Hardy"
## [11] "Ken Watanabe"
```

The summary description of h1 indicates that hwd is indeed a bipartite graph.
We can surmise that the ties link each actor to the movie that actor was in. The
description also reveals that the network has 1,365 nodes and 1,600 ties. This is a
little harder to decipher for an affiliation network, but given what we already know
we can figure out that there are 160 movie nodes, 1,205 actor nodes, and the 1,600
ties arise from each movie having links to just ten actors. (There are only 1,205
actors listed because some actors appear in more than one movie.)

The entire network is too large to show here, but we can examine a small subset
of it before doing more focused analyses. As a first step, we can take advantage of
igraph's ability to store plotting information within the network object itself. In
this case, the node color and shape can be designated by defining these as vertex
attributes. (Compare to how this was done in the previous plotting example, where
the node colors and shapes were designated within the plot function call.)

```
V(h1)$shape <- ifelse(V(h1)$type==TRUE,
                      "square","circle")
V(h1)$shape[1:10]
```

```
##  [1] "square" "square" "square" "square" "square"
##  [6] "square" "square" "square" "square" "square"
```

```
V(h1)$color <- ifelse(V(h1)$type==TRUE,
                      "red","lightblue")
```

For the first plot, we will look at a subset of Martin Scorsese movies that were
released in the past 15 years. This example also illustrates how to create a subgraph
by extracting only the edges that are incident to vertices with certain properties
(in this case the name matches one of the three listed Scorsese movies). The key
here is the inc special function of the E() edge iterator. The inc function takes
a vertex sequence as an argument, and returns the incident edges. In this case, we
are extracting all of the edges that are incident to the three Scorsese movies. For

more information see the `igraph` help entry on iterators, which can be found with `help(E)`. The resulting graphic highlights the special role of Leonardo DiCaprio in these Scorsese movies, being the only actor to star in all three (Fig. 9.4).

```
h2 <- subgraph.edges(h1, E(h1)[inc(V(h1)[name %in%
    c("The Wolf of Wall Street", "Gangs of New York",
    "The Departed")])])
plot(h2, layout = layout_with_kk)
```

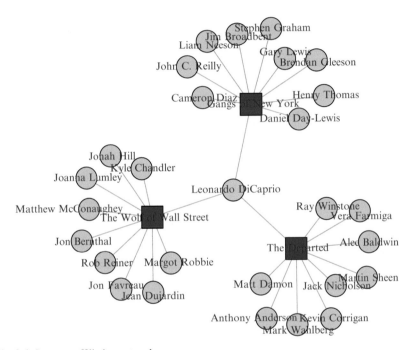

Fig. 9.4 Scorsese affiliation network

What can be learned from the entire Hollywood network? Most network descriptive statistics can be applied to affiliation networks, but they often need to be adjusted either in how they are constructed or how they are interpreted. For example, the overall density of the affiliation network can be easily calculated, but it is not very meaningful given how the network data were collected (every actor by definition is connected to a movie) and that there can be no ties among either the movie nodes or among the actor nodes.

```
graph.density(h1)
```

```
## [1] 0.00172
```

Instead, node degree may be more informative, at least for actors. (In this dataset, every movie has the same degree = 10). The `degree` function allows specification of which vertices to include, and that is used to select only the actors (for which the node characteristic `type` is FALSE).

```
table(degree(h1,v=V(h1)[type==FALSE]))
```

```
## 
##   1   2   3   4   5   6   7   8
## 955 165  47  23  11   2   1   1
```

```
mean(degree(h1,v=V(h1)[type==FALSE]))
```

```
## [1] 1.33
```

This shows that the vast majority of actors only appeared in one movie, but there were 15 actors who each starred in five or more movies since 1999. Across all the actors, they starred in an average of 1.3 movies. This information can then be used to identify the busiest actors of the past decade and a half. They owe a lot of thanks to Harry Potter and Batman!

```
V(h1)$deg <- degree(h1)
V(h1)[type==FALSE & deg > 4]$name
```

```
##  [1] "Leonardo DiCaprio"  "Emma Watson"
##  [3] "Richard Griffiths"  "Harry Melling"
##  [5] "Daniel Radcliffe"   "Rupert Grint"
##  [7] "James Franco"       "Ian McKellen"
##  [9] "Martin Freeman"     "Bradley Cooper"
## [11] "Christian Bale"     "Samuel L. Jackson"
## [13] "Natalie Portman"    "Brad Pitt"
## [15] "Liam Neeson"
```

```
busy_actor <- data.frame(cbind(
  Actor = V(h1)[type==FALSE & deg > 4]$name,
  Movies = V(h1)[type==FALSE & deg > 4]$deg
))
busy_actor[order(busy_actor$Movies,decreasing=TRUE),]
```

```
##                Actor Movies
## 5  Daniel Radcliffe       8
## 11   Christian Bale       7
## 1  Leonardo DiCaprio      6
## 2       Emma Watson       6
## 3  Richard Griffiths      5
## 4      Harry Melling      5
## 6       Rupert Grint      5
## 7       James Franco      5
```

```
## 8          Ian McKellen        5
## 9         Martin Freeman       5
## 10        Bradley Cooper       5
## 12  Samuel L. Jackson          5
## 13       Natalie Portman       5
## 14            Brad Pitt        5
## 15          Liam Neeson        5
```

This tells us who the busiest actors were. If we wanted to assess the popularity of the actors based on the popularity of the movies they appeared in, we could do this by accessing the characteristics of each movie that each actor starred in. This is slightly more complicated than the previous example, but can be done by utilizing igraph's abilities to identify the adjacent neighbors for any node in the graph.

The following code loops through the actor nodes in the network, and sums up the IMDBrating for all the neighbors of each node. Note that the loop only assigns the summed IMDBrating scores for the actor nodes (which are listed after the first 160 movie nodes).

```
for (i in 161:1365) {
  V(h1)[i]$totrating <- sum(V(h1)[nei(i)]$IMDBrating)
  }
```

Once we have this we can once again examine the most popular actors, which is based on both the number of movies, and the overall popularity of those movies.

```
max(V(h1)$totrating,na.rm=TRUE)
```

```
## [1] 60.9
```

```
pop_actor <- data.frame(cbind(
  Actor = V(h1)[type==FALSE & totrating > 40]$name,
  Popularity = V(h1)[type==FALSE &
                    totrating > 40]$totrating))
pop_actor[order(pop_actor$Popularity,decreasing=TRUE),]
```

```
##                   Actor Popularity
## 3   Daniel Radcliffe        60.9
## 4     Christian Bale        55.5
## 1  Leonardo DiCaprio        49.6
## 2      Emma Watson          45
## 5         Brad Pitt        40.5
```

Finally, network characteristics can always be examined using more traditional graphical and statistical approaches. For example, we can see if the busiest actors are starring in more popular movies, on average. First, we calculate an avgrating characteristic that is based on the mean IMDBrating, rather than the sum. Then, a simple scatterplot and regression are examined to see the relationship between number of movies and the average ratings of those movies. The results suggest that there is not a strong relationship between how busy an actor has been and the popularity of their movies. However, the scatterplot also suggests that actors who appear in less popular movies are most likely to appear in only one or two movies (Fig. 9.5).

```
for (i in 161:1365) {
  V(h1)[i]$avgrating <- mean(V(h1)[nei(i)]$IMDBrating)
  }
num <- V(h1)[type==FALSE]$deg
avgpop <- V(h1)[type==FALSE]$avgrating
summary(lm(avgpop ~ num))

  ##
  ## Call:
  ## lm(formula = avgpop ~ num)
  ##
  ## Residuals:
  ##     Min      1Q Median     3Q    Max
  ## -3.986 -0.433  0.198  0.617  1.614
  ##
  ## Coefficients:
  ##               Estimate Std. Error t value Pr(>|t|)
  ## (Intercept)    7.3387     0.0544  134.91   <2e-16
  ## num            0.0471     0.0353    1.34     0.18
  ##
  ## Residual standard error: 0.96 on 1203 df
  ## Multiple R-sq:  0.00148,Adjusted R-sq:  0.000653
  ## F-statistic: 1.79 on 1 and 1203 DF,  p-value: 0.182


scatter.smooth(num,avgpop,col="lightblue",
               ylim=c(2,10),span=.8,
               xlab="Number of Movies",
               ylab="Avg. Popularity")
```

9.3.2 Analysis of the Actor and Movie Projections

Following the same procedures as presented in Sect. 9.2.4, the two projections of the Hollywood affiliation network can be created and analyzed. This will produce an actor network where actors have ties if they starred together in the same movie, and a movie network where the movies are connected if they shared the same actors. The actor projection will thus have 1,205 nodes, and the movie projection network will have 160 nodes.

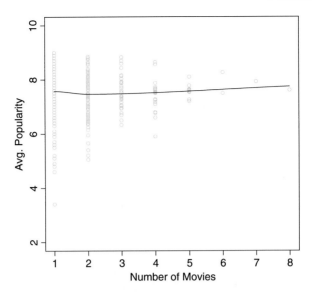

Fig. 9.5 Relationship between actor activity and popularity

```
h1.pr <- bipartite.projection(h1)
h1.act <- h1.pr$proj1
h1.mov <- h1.pr$proj2
h1.act

  ## IGRAPH UNW- 1205 6903 --
  ## + attr: name (v/c), year (v/n), IMDBrating
  ## | (v/n), MPAArating (v/c), shape (v/c),
  ## | color (v/c), deg (v/n), totrating (v/n),
  ## | avgrating (v/n), weight (e/n)
  ## + edges (vertex names):
  ## [1] Leonardo DiCaprio--Joseph Gordon-Levitt
  ## [2] Leonardo DiCaprio--Ellen Page
  ## [3] Leonardo DiCaprio--Tom Hardy
  ## [4] Leonardo DiCaprio--Ken Watanabe
  ## [5] Leonardo DiCaprio--Dileep Rao
  ## + ... omitted several edges

h1.mov

  ## IGRAPH UNW- 160 472 --
  ## + attr: name (v/c), year (v/n), IMDBrating
  ## | (v/n), MPAArating (v/c), shape (v/c),
  ## | color (v/c), deg (v/n), totrating (v/n),
  ## | avgrating (v/n), weight (e/n)
  ## + edges (vertex names):
```

```
## [1] Inception--The Wolf of Wall Street
## [2] Inception--Django Unchained
## [3] Inception--The Departed
## [4] Inception--Gangs of New York
## [5] Inception--Catch Me If You Can
## + ... omitted several edges
```

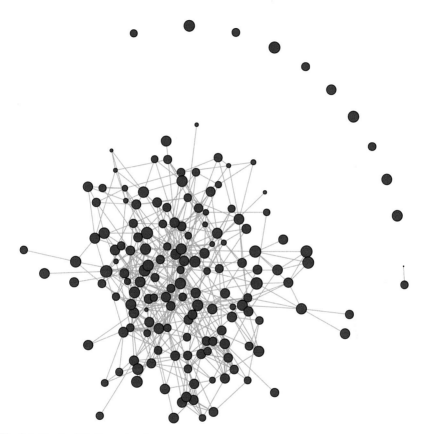

Fig. 9.6 Movie affiliation network

In this figure, the entire movie network is presented, with node size based on the
IMDBrating, so that more popular movies have larger nodes (Fig. 9.6).

```
op <- par(mar = rep(0, 4))
plot(h1.mov, vertex.color="red",
     vertex.shape="circle",
     vertex.size=(V(h1.mov)$IMDBrating)-3,
     vertex.label=NA)
```

```
par(op)
```

Some basic network descriptives provide more information about the Hollywood movie network. Although there are some isolated movies (i.e., movies that did not share actors with any of the other movies), most (148) of the movies form a large connected component.

```
graph.density(h1.mov)
```

```
## [1] 0.0371
```

```
no.clusters(h1.mov)
```

```
## [1] 12
```

```
clusters(h1.mov)$csize
```

```
## [1] 148    1    1    1    1    1    1    2    1    1    1
## [12]   1
```

```
table(E(h1.mov)$weight)
```

```
##
##    1    2    3    4    5    6    7   10
## 411   21   12   16    6    1    2    3
```

The complete movie network can be filtered to examine the single large connected component. In the next figure the edge width has been set to equal the square root of `weight` edge attribute. This results in the ties being thicker for movies that share more actors between them (Fig. 9.7).

```
h2.mov <- induced.subgraph(h1.mov,
              vids=clusters(h1.mov)$membership==1)
```

```
plot(h2.mov,vertex.color="red",
      edge.width=sqrt(E(h1.mov)$weight),
      vertex.shape="circle",
      vertex.size=(V(h2.mov)$IMDBrating)-3,
      vertex.label=NA)
```

The previous figure is still large and the relatively high density makes it somewhat challenging to interpret any interesting structural features. To help with that, we can identify the higher density cores of the graph, and use that to 'zoom in' on the more interconnected part of the network. (See Chap. 8 for more information.) This network is small enough that we can add node labels to help with the interpretation. This helps us see that the most tightly connected sections of the network

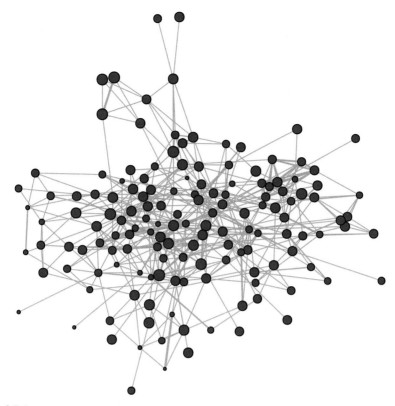

Fig. 9.7 Largest component of movie affiliation network

correspond to popular movie series, in particular Harry Potter, Batman, Star Wars, and The Hobbit. This makes sense because movies in a series will naturally share many or most of the same actors (Fig. 9.8).

```
table(graph.coreness(h2.mov))

  ##
  ##  1  2  3  4  5  6  7
  ## 11  5 23 65 29  7  8
```

```
h3.mov <- induced.subgraph(h2.mov,
              vids=graph.coreness(h2.mov)>4)
h3.mov

  ## IGRAPH UNW- 44 158 --
  ## + attr: name (v/c), year (v/n), IMDBrating
  ## | (v/n), MPAArating (v/c), shape (v/c),
  ## | color (v/c), deg (v/n), totrating (v/n),
```

```
##  | avgrating (v/n), weight (e/n)
##  + edges (vertex names):
##  [1] Inception--The Wolf of Wall Street
##  [2] Inception--Django Unchained
##  [3] Inception--The Dark Knight Rises
##  [4] Inception--The Dark Knight
##  [5] Inception--The Departed
##  + ... omitted several edges

plot(h3.mov,vertex.color="red",
     vertex.shape="circle",
   edge.width=sqrt(E(h1.mov)$weight),
   vertex.label.cex=0.7,vertex.label.color="darkgreen",
   vertex.label.dist=0.3,
    vertex.size=(V(h3.mov)$IMDBrating)-3)
```

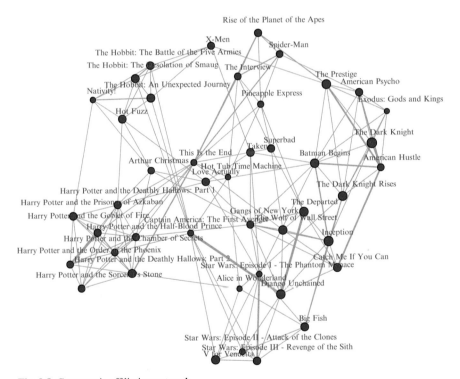

Fig. 9.8 Core movie affiliation network

Part IV
Modeling

Chapter 10
Random Network Models

> *What is this? A center for ants? How can we be expected to teach children to learn how to read... if they can't even fit inside the building? (Derek Zoolander, in the movie* Zoolander, *after looking at a miniature model of a school building.)*

10.1 The Role of Network Models

According to Linton Freeman (2004), modern social network analysis has four main characteristics:

1. It is motivated by a structural intuition based on ties linking social actors;
2. It is grounded in systematic empirical data;
3. It draws heavily on graphic imagery; and
4. It relies on the use of mathematical and/or computational models.

The preceding sections of this book focused on the first three elements of Freeman's characterization. The next four chapters now turn to consider his last point, the utility of modeling in network analysis. Scientific models are simplified descriptions of the real world that are used to predict or explain the characteristics or behavior of the phenomenon of interest. Models can be used in network science in the same way. With network models we can move beyond simple description to build and test hypotheses about network structures, formation processes, and network dynamics

In this chapter, a number of basic mathematical models of network structure and formation are covered. These are important models in the history of network science, but they are still useful today to provide insight into fundamental properties of social networks, to serve as baseline or comparison models for empirical social networks, and to act as building blocks for more complex network simulations.

Well over a dozen functions are provided in `igraph` for generating random networks based on a number of mathematical algorithms and heuristics. These all use 'game' as the final part of the function name, for example `barabasi.game()` produces scale-free random graphs based on the Barabási-Albert model (1999). In the rest of this chapter, a number of important mathematical network models available in `igraph` are presented, along with some examples of how to use these models to explore network properties and as comparisons to observed, empirical social networks.

© Springer International Publishing Switzerland 2015
D.A. Luke, *A User's Guide to Network Analysis in R*, Use R!,
DOI 10.1007/978-3-319-23883-8_10

10.2 Models of Network Structure and Formation

10.2.1 Erdős-Rényi Random Graph Model

The earliest historically, and still one of the most important mathematical models of network structure, is the random graph model first developed by Paul Erdős and Alfred Rényi in the late 1950s and early 1960s (Newman 2010). This is sometimes called the *Poisson random graph model* (because of the Poisson degree distribution of large random graphs), or sometimes even just the *random graph model*. The model is quite simple, $G(n,m)$, where a random graph G is defined with n vertices and m edges among those vertices chosen randomly. An equivalent model that is easier to work with is $G(n,p)$, where instead of specifying m edges, each edge appears in the graph with probability p. This random graph model is implemented in igraph with the erdos.reny.game() function. A random graph is produced by specifying the size of the desired network, and either the number of edges, or the probability of observing an edge. The type argument is used to specify whether the second argument should be interpreted as probability of an edge p, or number of edges m.

```
library(igraph)
g <- erdos.renyi.game(n=12,10,type='gnm')
g

  ## IGRAPH U--- 12 10 -- Erdos renyi (gnm) graph
  ## + attr: name (g/c), type (g/c), loops (g/l),
  ## | m (g/n)
  ## + edges:
  ##  [1] 4-- 5 3-- 6 2-- 8 1-- 9 8-- 9 8--10 1--11
  ##  [8] 6--11 8--12 9--12

graph.density(g)

  ## [1] 0.152
```

The random nature of the graphs can be seen by producing and examining multiple graphs. In each case the number of vertices is the same, but the ties are randomly determined (Fig. 10.1).

```
op <- par(mar=c(0,1,3,1),mfrow=c(1,2))
plot(erdos.renyi.game(n=12,10,type='gnm'),
                    vertex.color=2,
                    main="First random graph")
plot(erdos.renyi.game(n=12,10,type='gnm'),
                    vertex.color=4,
                    main="Second random graph")
par(op)
```

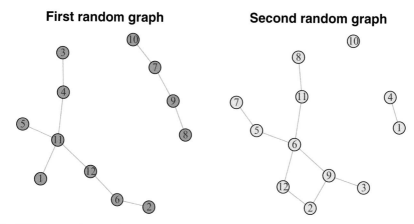

Fig. 10.1 Two random graphs

Despite the simplicity of the random graph model, it has led to a number of important discoveries about network structures. First, as suggested above, for large n the network will have a Poisson degree distribution (Fig. 10.2).

```
g <- erdos.renyi.game(n=1000,.005,type='gnp')
plot(degree.distribution(g),
     type="b",xlab="Degree",ylab="Proportion")
```

More unexpectedly, it turns out that random graphs become entirely connected for fairly low values of average degree. That means even when edges are determined randomly, each individual network member does not have to be connected to too many other members for the network itself to be connected (i.e., the network has only one component). More precisely, if p is greater than $\frac{ln}{n}$, then the random graph is likely to be connected in one large component (Newman 2010). The average degree of a random graph, c, is related to graph size and edge probability:

$$c = (n-1)p$$

So this means that across the range of network sizes typically seen in social network analysis (say, 100–10,000), the average degree required to have a completely connected network will be less than approximately 12. The following random graph simulation and plot demonstrates this relationship (Fig. 10.3).

```
crnd <- runif(500,1,8)
cmp_prp <- sapply(crnd,function(x)
             max(clusters(erdos.renyi.game(n=1000,
                p=x/999))$csize)/1000)
```

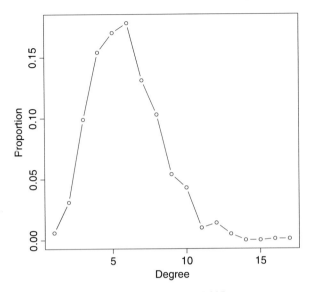

Fig. 10.2 Degree distribution for G with n = 1,000 and p = 0.005

```
smoothingSpline <- smooth.spline(crnd,cmp_prp,
                                 spar=0.25)
plot(crnd,cmp_prp,col='grey60',
     xlab="Avg. Degree",
     ylab="Largest Component Proportion")
lines(smoothingSpline,lwd=1.5)
```

   This demonstration requires some unpacking to easily understand. First, a vector is created with 500 random values ranging from one to eight. This will be used as input to a function that will create 500 random graphs, with average degree varying from 1 to 8. Next, the `sapply()` function is used to call the random graph function repeatedly, and assign the results to `cmp_prp`. The function itself creates each random graph with 1,000 nodes, and with p equal to the desired average degree divided by 999. (From above, $p = \frac{c}{n-1}$.) Once the random graph is produced, the size of the largest component is calculated using the `clusters()` function. This number is then divided by the size of the network (1,000) to get the proportion of the network accounted for by the largest component. The results show that for random networks of 1,000 nodes, the network will be almost or completely connected when the average degree is larger than four or five.

   Another surprising property of random graphs is that the connected random graphs are quite compact. That is, the diameter of the largest components in random graphs stays relatively small even for large networks.

```
n_vect <- rep(c(50,100,500,1000,5000),each=50)
g_diam <- sapply(n_vect,function(x)
```

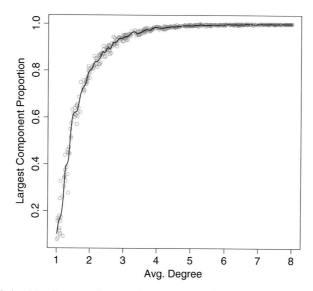

Fig. 10.3 Relationship of average degree and connectedness in random graphs

```
diameter(erdos.renyi.game(n=x,p=6/(x-1))))
```

```
library(lattice)
bwplot(g_diam ~ factor(n_vect), panel = panel.violin,
    xlab = "Network Size", ylab = "Diameter")
```

The above code runs a total of 250 simulations, producing random graphs from 50 to 5,000 nodes. As the plot shows, although the size of the graphs increases across two orders of magnitude, the diameter of the largest component in each graph increases much more slowly, from about five to ten (Fig. 10.4). These two characteristics of random graphs: being completely connected with low average degree, and the diameter increasing slowly relative to graph size, may be partly responsible for some of the 'small-world' characteristics of real-world social networks (Newman 2010).

10.2.2 Small-World Model

The Erdős-Rényi random graph model has one major limitation in that it does not describe the properties of many real-world social networks. In particular, fully random graphs have degree distributions that do not match observed networks very well, and they also have quite low levels of clustering (transitivity).

One type of model, called the small-world model by Watts and Strogatz (1998), produces random networks that are somewhat more realistic than Erdős-Rényi graphs. In particular, small-world model networks have more realistic levels of transitivity along with small diameters.

The small-world model starts with a circle of nodes, where each node is connected to its *c* immediate neighbors (forming a formal lattice structure). Then, a small number of existing edges are *rewired*, where they are removed and then replaced with another tie that connects two random nodes. If the rewiring probability is 0, then we end up with the original lattice network. When *p* is 1, then we have an Erdős-Rényi random graph. The main interesting discovery of Watts and Strogatz (and others), is that only a small fraction of ties needs to be rewired to dramatically reduce the diameter of the network.

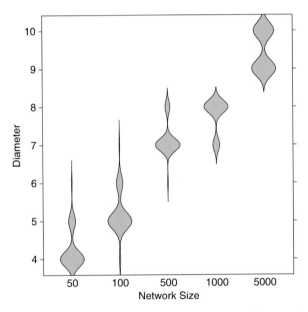

Fig. 10.4 Relationship of random graph size and diameter, for average degree $= 6$

Figure 10.5 shows how various small-world model networks look with different rewiring probabilities. The `watts.strogatz.game()` is called to produce a small-world network of 30 nodes. Setting the option `nei=2` (for neighborhood) will start the network with each node tied to the closest two neighbors on either side. This results in each node having degree $= 4$.

```
g1 <- watts.strogatz.game(dim=1, size=30, nei=2, p=0)
g2 <- watts.strogatz.game(dim=1, size=30, nei=2, p=.05)
g3 <- watts.strogatz.game(dim=1, size=30, nei=2, p=.20)
g4 <- watts.strogatz.game(dim=1, size=30, nei=2, p=1)
op <- par(mar=c(2,1,3,1),mfrow=c(2,2))
plot(g1,vertex.label=NA,layout=layout_with_kk,
```

```
    main=expression(paste(italic(p)," = 0")))
plot(g2,vertex.label=NA,
    main=expression(paste(italic(p)," = .05")))
plot(g3,vertex.label=NA,
    main=expression(paste(italic(p)," = .20")))
plot(g4,vertex.label=NA,
    main=expression(paste(italic(p)," = 1")))
par(op)
```

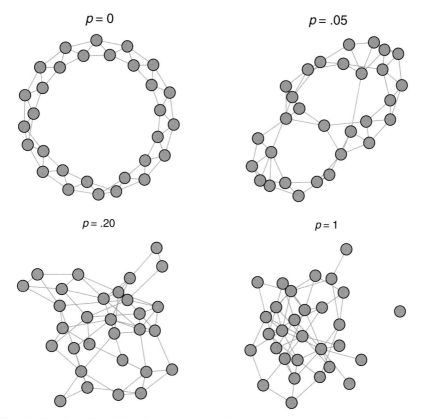

Fig. 10.5 Small-world models with increasing rewiring probabilities

The following simulation and figure shows how quickly rewiring reduces the diameter of a network in the small-world model. Working with a network with 100 nodes, each node starts out connected to its two neighbors on each side. The graph will thus have 200 edges. The starting diameter of the lattice network is 25 (getting from one node to the other side of the circle takes 25 steps).

```
g100 <- watts.strogatz.game(dim=1,size=100,nei=2,p=0)
g100
```

```
## IGRAPH U--- 100 200 -- Watts-Strogatz random grap
## + attr: name (g/c), dim (g/n), size (g/n),
## | nei (g/n), p (g/n), loops (g/l), multiple
## | (g/l)
## + edges:
##  [1]  1-- 2  2-- 3  3-- 4  4-- 5  5-- 6  6-- 7
##  [7]  7-- 8  8-- 9  9--10 10--11 11--12 12--13
## [13] 13--14 14--15 15--16 16--17 17--18 18--19
## [19] 19--20 20--21 21--22 22--23 23--24 24--25
## [25] 25--26 26--27 27--28 28--29 29--30 30--31
## [31] 31--32 32--33 33--34 34--35 35--36 36--37
## + ... omitted several edges
```

```
diameter(g100)
```

```
## [1] 25
```

The simulation is set to calculate 300 networks, ten each for the number of edges to rewire ranging from 1 to 30. Because we know how many edges are in each graph (200), the rewiring probability can be calculated by the number of rewired edges divided by total number of edges. If 30 edges are rewired, then, the probability is 0.15.

```
p_vect <- rep(1:30,each=10)
g_diam <- sapply(p_vect,function(x)
    diameter(watts.strogatz.game(dim=1, size=100,
                               nei=2, p=x/200)))
smoothingSpline = smooth.spline(p_vect, g_diam,
                             spar=0.35)
plot(jitter(p_vect,1),g_diam,col='grey60',
    xlab="Number of Rewired Edges",
    ylab="Diameter")
lines(smoothingSpline,lwd=1.5)
```

The plot demonstrates that after only rewiring ten of the edges ($p = 0.05$), the diameter has shrunk at least 60 %, from 25 to about 10 (Fig. 10.6).

10.2.3 Scale-Free Models

An important limitation of the previous two mathematical network models is that they produce graphs with degree distributions that are not representative of many

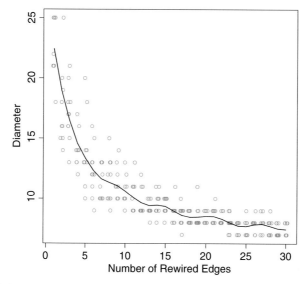

Fig. 10.6 Relationship of rewiring probability to network diameter for the small-world model

real-world social networks. Numerous studies, in fact, have shown that a wide variety of observed networks have heavy-tailed degree distributions that approximately follow a power law. These are typically called *scale-free* networks. For example, both the network of sexual partners and the World-Wide-Web exhibit this scale-free pattern (Broder et al. 2000; Liljeros et al. 2001). That is, some people have many sexual partners (high degree), but most people have a small number of sexual partners. Similarly, some websites have a very large number of other websites connected to them, but most websites have only a few connections.

How does this power-law characteristic feature of scale-free social networks arise? A number of network scientists have explored this question, and have determined that a network formation process of *cumulative advantage*, or *preferential attachment* can explain this. That is, as networks grow, new nodes are more likely to form ties with other nodes that already have many ties, due to their visibility in the network. This 'rich-gets-richer' phenomena has been shown to lead to the power-law distribution in networks (de Solla Price 1976; Barabási and Albert 1999).

The preferential attachment model of Barabási and Albert is implemented in igraph with the barabasi.game() function. This is a more complicated algorithm than those for the previous models, partly because this is a network growth model, not just a static network structure model.

Figure 10.7 displays a 500-node network that is formed with this preferential attachment model. The default behavior of the algorithm is that as each new node is added to the network, it is connected to another node in the network, with probability proportional to the degree of that node. Thus, some nodes in the network will end up with many more ties than most of the other nodes. The code for this example highlights these *hubs* by coloring the nodes with degree > 9 red. Also, the nodes are sized based on their degree, using the `rescale()` function from Chap. 5.

```
g <- barabasi.game(500, directed = FALSE)
V(g)$color <- "lightblue"
V(g)[degree(g) > 9]$color <- "red"
node_size <- rescale(node_char = degree(g), low = 2,
    high = 8)
plot(g, vertex.label = NA, vertex.size = node_size)
```

We can see the heavy-tail distribution in a few ways. The median degree is 1, and the mean is close to 2. The highest degree is 27. The pattern is easily seen in Fig. 10.8. The left panel shows the raw degree distribution, while the right panel displays the same degree distribution, but on log-scales. If the distribution follows a power-law, then the datapoints should fall along a straight line (at least for the tail). (Note that it is difficult to assess power-law relationships with small networks.)

```
median(degree(g))
```

```
## [1] 1
```

```
mean(degree(g))
```

```
## [1] 2
```

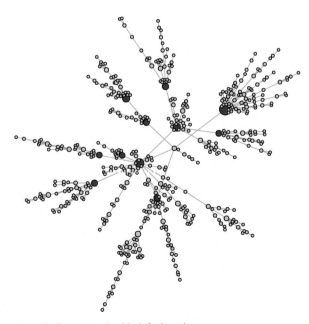

Fig. 10.7 Example scale-free network with default options

```
table(degree(g))

## 
##    1    2    3    4    5    6    7    9   10   11   12   14
## 314   86   41   23   11   10    3    1    3    2    2    1
##   16   19   27
##    1    1    1
```

```
op <- par(mfrow=c(1,2))
plot(degree.distribution(g),xlab="Degree",
    ylab="Proportion")
plot(degree.distribution(g),log='xy',
    xlab="Degree",ylab="Proportion")
par(op)
```

The user can adjust a number of parameters in the `barabasi.game()` function to produce a wide variety of preferential attachment networks. For example, the following code produces a network that might be viewed as a little more realistic than that shown in the previous figure. Here, instead of each new node connecting to exactly one other node, the `out.dist` option is used to specify a distribution of tie probabilities. In this case a new node will have a tie to 0 other nodes (isolate) 25 % of the time, will be tied to one other node 50 % of the time, and to two nodes 25 % of

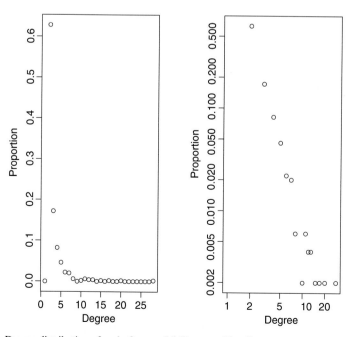

Fig. 10.8 Degree distribution of scale-free model (linear and log-linear scales)

the time. Also, the `zero.appeal` option is used to make it somewhat more likely that new nodes are connected to previously existing isolates. Figure 10.9 shows that the resulting graph has produced a number of isolates and slightly fewer nodes with high degree.

```
g <- barabasi.game(500, out.dist = c(0.25, 0.5, 0.25),
    directed = FALSE, zero.appeal = 1)
V(g)$color <- "lightblue"
V(g)[degree(g) > 9]$color <- "red"
node_size <- rescale(node_char = degree(g), low = 2,
    high = 8)
plot(g, vertex.label = NA, vertex.size = node_size)
```

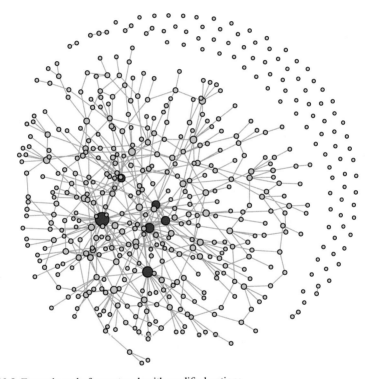

Fig. 10.9 Example scale-free network with modified options

Finally, to illustrate how preferential attachment networks grow, the following figure shows what the networks look like at four different stages, with graph sizes of 10, 25, 50, and 100 nodes (Fig. 10.10).

```
g1 <- barabasi.game(10,m=1,directed=FALSE)
g2 <- barabasi.game(25,m=1,directed=FALSE)
g3 <- barabasi.game(50,m=1,directed=FALSE)
```

```
g4 <- barabasi.game(100,m=1,directed=FALSE)

op <- par(mfrow=c(2,2),mar=c(4,0,1,0))
plot(g1, vertex.label= NA, vertex.size = 3,
    xlab = "n = 10")
plot(g2, vertex.label= NA, vertex.size = 3,
    xlab = "n = 25")
plot(g3, vertex.label= NA, vertex.size = 3,
    xlab = "n = 50")
plot(g4, vertex.label= NA, vertex.size = 3,
    xlab = "n = 100")
par(op)
```

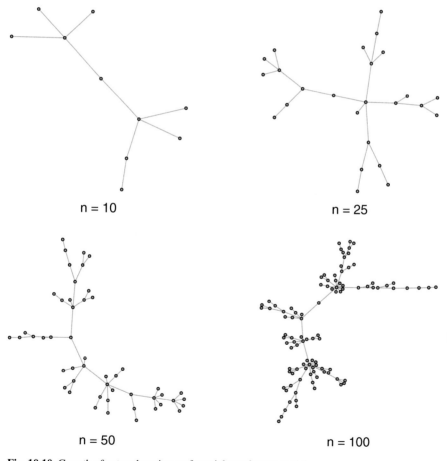

Fig. 10.10 Growth of networks using preferential attachment model

10.3 Comparing Random Models to Empirical Networks

The mathematical models described here, as well as many others, have been used
to study the theoretical properties of networks. These models can also be useful as
comparisons to empirical social networks. As as simple example of this, we can
explore some basic network properties of the lhds data taken from Harris' 2013
book on exponential random graph models (see Chap. 11). The lhds network is
made up of communication ties among 1,283 leaders of local public health depart-
ments. The network has quite low density, although the average degree is over four
(Fig. 10.11).

```
data(lhds)
ilhds <- asIgraph(lhds)
ilhds
```

```
## IGRAPH U--- 1283 2708 --
## + attr: title (g/c), hivscreen (v/c), na
## | (v/l), nutrition (v/c), popmil (v/n),
## | state (v/c), vertex.names (v/c), years
## | (v/n), na (e/l)
## + edges:
## [1]   2--  10  2--  11  2--  19  2--  20  5--1003
## [6]   5--   6  6--  11  6--  17 10--  11 11--  19
## [11] 11--  26  2--  12  6--  12 10--  12 11--  12
## [16] 12--  19 12--  26  9--  14 14--  15 14--  18
## [21] 14--  25 14--  27 14-- 226 14-- 414 14-- 697
## + ... omitted several edges
```

```
graph.density(ilhds)
```

```
## [1] 0.00329
```

```
mean(degree(ilhds))
```

```
## [1] 4.22
```

The following code builds three network models that have the same size and
approximately the same density as the lhds network. By comparing the characteristics of the network models with the empirical network, we can highlight the
interesting or important characteristics of the empirical network that might be worth
further exploration (by using, for example, the types of statistical modeling and simulation approaches presented in the next few chapters).

```
g_rnd <- erdos.renyi.game(1283,.0033,type='gnp')
g_smwrld <- watts.strogatz.game(dim=1,size=1283,
                                 nei=2,p=.25)
```

```
g_prfatt <- barabasi.game(1283,out.dist=c(.15,.6,.25),
                          directed=FALSE,zero.appeal=2)
```

Table 10.1 presents some descriptive network statistics for the three models and the `lhds` network. Although each model captures some of the characteristics of the observed network, none of them match across all the statistics. In particular, the `lhds` network has much higher transitivity than any of the models (Fig. 10.12).

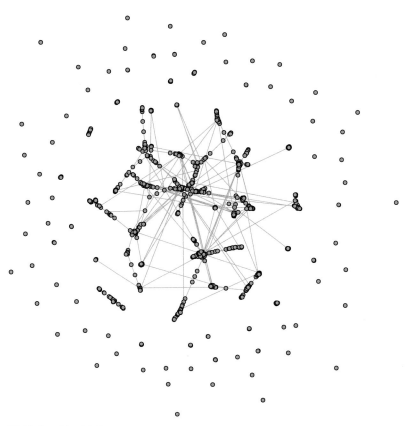

Fig. 10.11 Local health department communication network

Name	Size	Density	Avg. degree	Transitivity	Isolates
Erdos-Renyi	1283	0.003	4.404	0.002	21
Small world	1283	0.003	4.000	0.088	1
Preferential attachment	1283	0.002	2.195	0.003	109
Health department	1283	0.003	4.221	0.306	58

Table 10.1 Comparison of model and empirical network characteristics

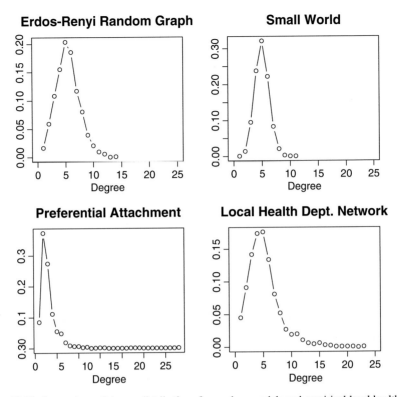

Fig. 10.12 Comparison of degree distributions for random models and empirical local health department network

Chapter 11
Statistical Network Models

> *Prediction and explanation are exactly symmetrical.*
> *Explanations are, in effect, predictions about what has*
> *happened; predictions are explanations about what's going to*
> *happen.* (John Rogers Searle)

11.1 Introduction

As suggested in the previous chapter, for most of the history of network science analysts were limited to network visualization and network description. Only in the past couple of decades has network statistical theory and computational power developed enough to allow for valid and feasible statistical modeling of networks. The primary barrier to statistical network modeling was the fundamental assumption of independence of observations that underlies much of traditional statistical theory. Networks by definition are non-independent. If you know that one actor is tied to another actor, you have information about the second actor that is dependent on the first.

Over the years statistical theorists gradually developed increasingly sophisticated models that could be applied to empirical network data, including dyadic dependence and dyadic independence models (such as p^*, see Harris 2013). The focus of this chapter is on *exponential random graph models* (ERGMs), which have turned out to be the most powerful, flexible, and widely used modeling approach for building and testing statistical models of networks.

An ERGM is a true generative statistical model of network structure and characteristics (Hunter et al. 2008). This means that inferential hypotheses can be proposed and tested. It is generative in the sense that characteristics of the individual elements in the network (i.e., actors) and local structural properties can be used to predict properties of the entire network (e.g., diameter, degree distribution, etc.). ERGMs are popular for at least four reasons. First, they can handle the complex dependencies of network data without the types of degeneracy problems that were frequently encountered in earlier network models. Second, ERGMs are flexible and can handle many different types of predictors and covariates. Third, the generative approach where overall network characteristics are predicted from individual actor and local structural properties enhances the validity of the models. Finally, ERGM

© Springer International Publishing Switzerland 2015
D.A. Luke, *A User's Guide to Network Analysis in R*, Use R!,
DOI 10.1007/978-3-319-23883-8_11

models have been implemented in programming suites and statistical packages such as R, making it easier for applied analysts to build, test, and disseminate the results of their network models.

ERGMs are fit using Monte Carlo Markov Chain maximum-likelihood estimation (MCMC). The estimates of the statistical parameters are based on an underlying simulation, where many (typically thousands) of networks are produced to reflect the particular model being tested. One way to express the basic ERGM model is:

$$P(y_{ij} = 1 \mid Y_{ij}^C) = \left(\frac{1}{c}\right) exp\{\sum_{k=1}^{K} \theta_k z_k(y)\}$$

This shows that the model is predicting the probability of a tie between actors i and j, conditional on the rest of the network (all other ties). The thetas (θ_k) are the coefficients of the network statistics of interest, one for each of the K included statistics, $z_k(y)$. $\left(\frac{1}{c}\right)$ is simply a normalizing constant that ensures that the probabilities stay within 0 and 1. (See Harris 2013, for more details.)

ERGMs are implemented in the ergm package that is contained in the statnet suite of network analysis packages in R. It is actively maintained and developed by network scientists at the University of Washington, among others. More technical details of ergm are provided in excellent papers by Hunter and colleagues (2008) as well as Goodreau (2007).

As suggested above, one of the strengths of ERGMs is the ability to handle a wide variety of predictors. In fact, the documentation for ergm lists more than a hundred possible types of terms that can be included in an ergm model specification. It is easier to navigate the possibilities once one knows that all the possible predictors in an ERGM fall into one of four broad categories: node-level predictors, dyadic predictors, relational predictors, and local structural predictors.

Table 11.1 lists these categories, along with some of the most commonly used ergm terms for each type. The first type of predictor is node or actor-level characteristics, where having a particular characteristic is hypothesized to affect the likelihood of observing a tie. For example, if you hypothesize that girls are more likely to make friends than boys in middle school, then you could use actor gender as a node-level predictor. Dyad-level predictors are used when you hypothesize that the characteristics of both actors in a dyad may influence the probability of observing a tie between those two actors. These types of predictors allow you to test hypotheses of assortative or disassortative mixing in networks, leading to patterns of homophily or heterophily. For example, if you think that friendships are more likely to be formed within the same grade in a middle school (assortative mixing), then you could use grade as a dyad-level predictor. The third type of predictor in ERGMs is a powerful option to use information about other relationships or ties when predicting the observed ties in a network. That is, you can use one type of network tie to predict a second type of tie (as long as they are both collected on the same set of network members). Finally, information about local structural properties of the observed network can be used as model covariates. This, for example, allows the network model to be conditioned on the observed degree distribution, or on the level of transitivity (closed triangles) that is observed.

Predictor type	Term
Node	nodefactor
	nodecov
Dyad	nodemix
	nodematch
	absdiff
Relation	edgecov
Structure	gwdegree
	gwdsp
	gwesp

Table 11.1 Common ERGM terms

In the following examples, we will build a series of ERGMs that use predictors from each of these four broad types of `ergm` terms. For more detailed information about the terms included in `ergm`, see `help('ergm-terms')`. A more general overview is provided by Morris, Handcock, and Hunter (2008). Finally, `ergm` includes a helpful vignette that shows how the various terms are related to one another (`vignette('ergm-term-crossRef')`).

11.2 Building Exponential Random Graph Models

To explore a number of the stochastic modeling possibilities of the `ergm` package, we will use network data that describe interorganizational relationships among 25 agencies within the Indiana state tobacco control program in 2010. These data include three different types of interorganizational ties: frequency of contact, level of collaboration, and whether each pair of agencies communicated with one another about a particular evidence-based guideline published by the Center's for Disease Control and Prevention (CDC), called *Best Practices for Tobacco Control*. This latter relationship is conceptualized as a type of dissemination tie. The following example models will focus on predicting the pattern of these dissemination ties among the Indiana tobacco control organizations.

The data are included in the `UserNetR` package as a network list object called `TCnetworks`. The network data include a number of node characteristics (e.g., `tob_yrs`, which records how long an agency has been working in tobacco control), edge characteristics, and a sociomatrix (`TCdist`) which contains the geographic distance between each pair of agencies. For more information, see the help file for `TCnetworks`.

Prior to modeling, the network data need to be extracted from the list object, and any preliminary descriptive and visualization tasks can be conducted.

```
data(TCnetworks)
TCcnt <- TCnetworks$TCcnt
TCcoll <- TCnetworks$TCcoll
```

```
TCdiss <- TCnetworks$TCdiss
TCdist <- TCnetworks$TCdist
summary(TCdiss,print.adj=FALSE)
```

```
## Network attributes:
##   vertices = 25
##   directed = FALSE
##   hyper = FALSE
##   loops = FALSE
##   multiple = FALSE
##   bipartite = FALSE
##   title = IN_Diffusion
## total edges = 103
##     missing edges = 0
##     non-missing edges = 103
## density = 0.343
##
## Vertex attributes:
##
## agency_cat:
##     numeric valued attribute
##     attribute summary:
##    Min. 1st Qu.  Median   Mean 3rd Qu.    Max.
##    1.00    2.00    2.00   3.24    5.00    6.00
##
## agency_lvl:
##     numeric valued attribute
##     attribute summary:
##    Min. 1st Qu.  Median   Mean 3rd Qu.    Max.
##    1.00    1.00    2.00   2.04    3.00    3.00
##
## lead_agency:
##     numeric valued attribute
##     attribute summary:
##    Min. 1st Qu.  Median   Mean 3rd Qu.    Max.
##    0.00    0.00    0.00   0.04    0.00    1.00
##
## tob_yrs:
##     numeric valued attribute
##     attribute summary:
##    Min. 1st Qu.  Median   Mean 3rd Qu.    Max.
##    1.00    3.00    4.50   6.76    9.00   21.00
##     vertex.names:
##     character valued attribute
##     25 valid vertex names
```

```
##
## No edge attributes
```

A quick examination of the network reveals that it is made up of three types of organizations (local, state, and national), is made up of one connected component that is fairly densely connected, and there is some variability of centrality across the network members (shown both by the betweenness centralization score, as well as the variability of the sizes of nodes in the figure, based on degree) (Fig. 11.1).

```
components(TCdiss)

   ## [1] 1

gden(TCdiss)

   ## [1] 0.343

centralization(TCdiss,betweenness,mode='graph')

   ## [1] 0.381
```

```
deg <- degree(TCdiss,gmode='graph')
lvl <- TCdiss %v% 'agency_lvl'
plot(TCdiss,usearrows=FALSE,displaylabels=TRUE,
        vertex.cex=log(deg),
        vertex.col=lvl+1,
        label.pos=3,label.cex=.7,
        edge.lwd=0.5,edge.col="grey75")
legend("bottomleft",legend=c("Local","State",
                            "National"),
        col=2:4,pch=19,pt.cex=1.5)
```

11.2.1 Building a Null Model

It is often useful to start the modeling process by building a null model, one with no substantive or structural predictors. This can be used as a baseline model to judge how much subsequent models are improving. A null model typically only has one term, `edges`, which produces a random graph model that has the same number of edges as the observed network.

Fitting an `ergm` model uses syntax similar to other statistical modeling functions in R. The `ergm` function is called with a model formula. This formula lists the observed network as the dependent variable, and then all of the `ergm` model terms are listed on the right hand side. Other options can also be set by the user. The `control` option is used to pass control parameters to the `ergm` algorithm. Here we

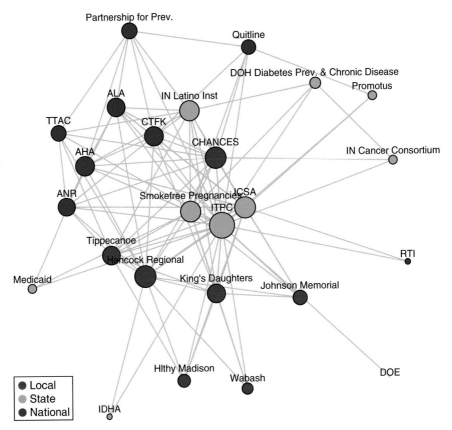

Fig. 11.1 Indiana tobacco control dissemination network

set the random number seed to ensure that the same results will be seen across multiple runs. The results of fitting the model are stored in a model object for further examination and analysis.

```
library(ergm)
DSmod0 <- ergm(TCdiss ~ edges,
              control=control.ergm(seed=40))
class(DSmod0)

  ## [1] "ergm"

summary(DSmod0)

  ##
  ## ==========================
  ## Summary of model fit
  ## ==========================
```

```
##
## Formula:    TCdiss ~ edges
##
## Iterations:   4 out of 20
##
## Monte Carlo MLE Results:
##          Estimate Std. Error MCMC % p-value
## edges     -0.648       0.122      0  <1e-04
##
##          Null Deviance: 416   on 300   degrees of freedom
##    Residual Deviance: 386   on 299   degrees of freedom
##
## AIC: 388     BIC: 392      (Smaller is better.)
```

A null model includes only the edges term. This acts as a type of intercept for the model, and ensures that the simulated networks have the same number of edges as the observed network. This can be seen by taking the logistic transformation of the edges parameter, which gives the overall density of network. This demonstrates that the null model is constrained by the number of edges in the observed network.

```
plogis(coef(DSmod0))
```

```
## edges
## 0.343
```

11.2.2 Including Node Attributes

Once a null model is obtained, more interesting models can be fit using a wide variety of predictors. Following the order suggested by Table 11.1, we start with main effect terms based on individual node characteristics. In the Indiana network, we know which agency is the lead agency (receiving funding from CDC), and we also know how long each agency has been working in tobacco control. It might be reasonable to assume that agencies are more likely to be connected to the lead agency. It also would make sense that agencies with a longer history of tobacco control experience would be more likely to be connected to other agencies. The following scatterplot does suggest that there may be a relationship between experience and interorganizational connections (Fig. 11.2).

```
scatter.smooth(TCdiss %v% 'tob_yrs',
               degree(TCdiss,gmode='graph'),
               xlab='Years of Tobacco Experience',
               ylab='Degree')
```

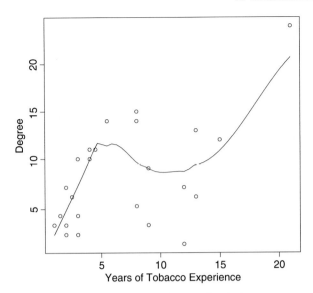

Fig. 11.2 Basic association between years of experience and node degree

Both of the above hypotheses can be formally tested. There are two main
`ergm` model terms for testing node characteristic main effects, `nodefactor` and
`nodecov`. `nodefactor` is used for a categorical attribute (e.g., `lead_agency`),
while `nodecov` is used for quantitative characteristics (such as `tob_yrs`).

```
DSmod1 <- ergm(TCdiss ~ edges +
               nodefactor('lead_agency') +
               nodecov('tob_yrs') ,
               control=control.ergm(seed=40))
summary(DSmod1)

   ##
   ## ==========================
   ## Summary of model fit
   ## --------------------------
   ##
   ## Formula:   TCdiss ~ edges +
   ##            nodefactor("lead_agency") +
   ##            nodecov("tob_yrs")
   ##
   ## Iterations:  16 out of 20
   ##
   ## Monte Carlo MLE Results:
   ##                          Estimate Std. Error
   ## edges                     -1.6783     0.3293
```

```
## nodefactor.lead_agency.1   17.9366     933.5551
## nodecov.tob_yrs                0.0599       0.0228
##                            MCMC % p-value
## edges                           0   <1e-04
## nodefactor.lead_agency.1        0   0.9847
## nodecov.tob_yrs                 0   0.0091
##
##        Null Deviance: 416   on 300   degrees of freedom
##    Residual Deviance: 323   on 297   degrees of freedom
##
## AIC: 329     BIC: 341     (Smaller is better.)
```

The results show a number of interesting points. First, the number of years in tobacco control is positively and significantly associated with the likelihood of observing a tie between the two agencies (remember that here a tie reflects a dissemination connection between two organizations). Second, although the parameter for the effect of being the lead agency is quite large, it is not significant. This appears to be due to low precision in the estimate. With only one lead agency in a small network of 25 members, there may not be the power to detect this effect. Finally, the AIC (Akaike information criterion) for this model with two predictors is lower than the AIC for the null model. This shows that this model is doing a better job of explaining the data than the baseline model.

Similar to logistic regression analysis, we can estimate the probability of observing certain types of ties using the fitted parameter estimates. This requires using the logistic transformation to obtain numbers that can be properly interpreted as probabilities. For example, based on Model 1, the following code calculates the probability that there is a dissemination tie between two agencies, one with 5 years of tobacco control experience, the other with 10 years of experience, neither of whom are the lead agency.

```
p_edg <- coef(DSmod1)[1]
p_yrs <- coef(DSmod1)[3]
plogis(p_edg + 5*p_yrs + 10*p_yrs)

  ## edges
  ## 0.314
```

The result is 0.31, which is just a little less than the overall density (i.e., overall probability of observing a tie) of the dissemination network.

11.2.3 Including Dyadic Predictors

A rich source of hypotheses for network structures derive from questions about homophily and heterophily. That is, are ties more or less likely between network

members who are similar to each other on some characteristic (homophily) or
dissimilar (heterophily). This is a type of dyadic interaction predictor, and `ergm`
includes a number of these terms.

The raw frequencies of observed ties among and between different types of actors
in the network can be displayed with the `mixingmatrix()` function. Here, for
example, we see that the most frequent dissemination ties (24) are observed between
local and state-level agencies (see `?TCnetworks` for the covariate code defini-
tions). It can be somewhat complicated to interpret these raw frequency patterns;
however, they can generate hypotheses about dyadic interrelationships that can be
formally tested in the ERGM.

```
mixingmatrix(TCdiss,'agency_lvl')

   ## Note:  Marginal totals can be misleading
   ##   for undirected mixing matrices.
   ##     1  2  3
   ## 1 13 24 14
   ## 2 24 16 23
   ## 3 14 23 13

mixingmatrix(TCdiss,'agency_cat')

   ## Note:  Marginal totals can be misleading
   ##   for undirected mixing matrices.
   ##      1  2  3  4  5  6
   ## 1  0 12  3  2  3  4
   ## 2 12 19 18  6  1 14
   ## 3  3 18  0  4  3  2
   ## 4  2  6  4  1  1  6
   ## 5  3  1  3  1  0  0
   ## 6  4 14  2  6  0  4
```

In the following models, the non-significant lead agency predictor is dropped.
Three version of Model 2 are estimated, showing different options for including the
dyadic comparison of agency level as a predictor.

```
DSmod2a <- ergm(TCdiss ~ edges +
            nodecov('tob_yrs') +
            nodematch('agency_lvl'),
            control=control.ergm(seed=40))
summary(DSmod2a)

   ##
   ## ===========================
   ## Summary of model fit
   ## ===========================
   ##
```

```
## Formula:    TCdiss ~ edges +
##                nodecov("tob_yrs") +
##                nodematch("agency_lvl")
##
## Iterations:  4 out of 20
##
## Monte Carlo MLE Results:
##                            Estimate Std. Error MCMC %
## edges                       -2.4808    0.3413      0
## nodecov.tob_yrs              0.1133    0.0201      0
## nodematch.agency_lvl         0.6875    0.2770      0
##                            p-value
## edges                       <1e-04
## nodecov.tob_yrs             <1e-04
## nodematch.agency_lvl         0.014
##
##        Null Deviance: 416  on 300  degrees of freedom
##    Residual Deviance: 342  on 297  degrees of freedom
##
## AIC: 348    BIC: 359    (Smaller is better.)
```

Model 2a uses the basic nodematch term to include one network predictor
that assesses the effect on the likelihood of a dissemination tie when both organizations are the same level (e.g., both are local organizations). This is a homophily
hypothesis, where we are testing if the same types of organizations are more likely
to communicate with one another. The positive and significant parameter indicates
that there is a homophily effect here.

```
DSmod2b <- ergm(TCdiss ~ edges +
            nodecov('tob_yrs') +
            nodematch('agency_lvl',diff=TRUE),
            control=control.ergm(seed=40))
summary(DSmod2b)

  ##
  ## ==========================
  ## Summary of model fit
  ## ==========================
  ##
  ## Formula:    TCdiss ~ edges +
  ##                nodecov("tob_yrs") +
  ##                nodematch("agency_lvl",
  ##       diff = TRUE)
  ##
  ## Iterations:  4 out of 20
```

```
##
## Monte Carlo MLE Results:
##                           Estimate Std. Error MCMC %
## edges                      -2.7792    0.3685       0
## nodecov.tob_yrs             0.1331    0.0217       0
## nodematch.agency_lvl.1      1.6145    0.4983       0
## nodematch.agency_lvl.2     -0.2148    0.3974       0
## nodematch.agency_lvl.3      1.3016    0.4422       0
##                           p-value
## edges                      <1e-04
## nodecov.tob_yrs            <1e-04
## nodematch.agency_lvl.1     0.0013
## nodematch.agency_lvl.2     0.5891
## nodematch.agency_lvl.3     0.0035
##
##        Null Deviance: 416  on 300  degrees of freedom
##    Residual Deviance: 330  on 295  degrees of freedom
##
## AIC: 340    BIC: 358    (Smaller is better.)
```

Model 2b shows how to test a hypothesis of differential homophily. Here, instead of one dyad term, there are three; one each for the three levels of the agency level characteristic. Thus, this model now has three homophily terms, one for local agencies, one for state, and one for national. The results suggest that the overall homophily effect is seen mainly at the local and national levels.

```
DSmod2c <- ergm(TCdiss ~ edges +
            nodecov('tob_yrs') +
            nodemix('agency_lvl',base=1),
            control=control.ergm(seed=40))
summary(DSmod2c)

    ##
    ## =========================
    ## Summary of model fit
    ## =========================
    ##
    ## Formula:   TCdiss ~ edges +
    ##            nodecov("tob_yrs") +
    ##            nodemix("agency_lvl", base = 1)
    ##
    ## Iterations:  4 out of 20
    ##
    ## Monte Carlo MLE Results:
    ##                      Estimate Std. Error MCMC %
    ## edges                 -1.1757     0.5372      0
```

```
## nodecov.tob_yrs        0.1340      0.0222       0
## mix.agency_lvl.1.2     -1.5805     0.5501       0
## mix.agency_lvl.2.2     -1.8354     0.6028       0
## mix.agency_lvl.1.3     -1.5363     0.5659       0
## mix.agency_lvl.2.3     -1.7073     0.5445       0
## mix.agency_lvl.3.3     -0.3110     0.6119       0
##                        p-value
## edges                  0.0294
## nodecov.tob_yrs        <1e-04
## mix.agency_lvl.1.2     0.0044
## mix.agency_lvl.2.2     0.0025
## mix.agency_lvl.1.3     0.0070
## mix.agency_lvl.2.3     0.0019
## mix.agency_lvl.3.3     0.6116
##
##      Null Deviance: 416  on 300  degrees of freedom
##  Residual Deviance: 330  on 293  degrees of freedom
##
## AIC: 344    BIC: 370    (Smaller is better.)
```

Finally, Model 2c shows how to include the most detailed tests for homophily and heterophily. The nodemix term includes dyadic comparisons for all the possible patterns of a categorical node attribute. The base option sets a reference category for the effects, otherwise all possible effects are included (which may lead to model stability problems). Here the reference category is (1,1), indicating the ties among the local-level agencies. Note that for smaller networks and categorical attributes with a large number of values, the mixing matrix may have empty cells (as well as use up many more degrees of freedom). These can cause problems for the model estimation and interpretation.

11.2.4 Including Relational Terms (Network Predictors)

The third type of predictor that can be included in ERGMs is a relational predictor, where information about ties among the network members is used to predict the likelihood of the dependent variable tie. This means that either other network variables can be used as predictors, or any other type of relational information among the actors.

In this example, we use both a traditional network predictor and a relational quantitative variable as predictors. First, we have the contact network variable, which measured the frequency of contact among the Indiana tobacco control agencies (1 = yearly; 2 = quarterly; 3 = monthly; 4 = weekly; and 5 = daily). Second, the physical distance (in miles) between each pair of agencies was calculated and stored in a statnet network object. For each of these predictors, the hypothesis is fairly evident. We expect that agencies who have more frequent contact with each other will be more likely share information on the *Best Practices* guidelines. Second, we also expect that the further apart two agencies are, the less likely they will be to disseminate information to each other.

To include relational quantitative predictors in an ERGM, the `edgecov` term can be used. Also, since these predictors are in the form of valued networks, the `attr` option points to the edge attribute that contains the appropriate quantitative information. First, we view a subset of the relational information for each of the predictors, then the ERGM is fit.

```
as.sociomatrix(TCdist,attrname = 'distance')[1:5,1:5]

##              1        2      3      4        5
## 1        0.00     1.94    492   1870     1.27
## 2        1.94     0.00    492   1869     1.91
## 3      491.87   491.97      0   2325   493.08
## 4     1869.88  1868.98   2325      0  1868.61
## 5        1.27     1.91    493   1869     0.00

as.sociomatrix(TCcnt,attrname = 'contact')[1:5,1:5]

##            ITPC Promotus RTI Quitline IDHA
## ITPC          0        5   4        4    3
## Promotus      5        0   3        4    0
## RTI           4        3   0        2    0
## Quitline      4        4   2        0    0
## IDHA          3        0   0        0    0

DSmod3 <- ergm(TCdiss ~ edges +
               nodecov('tob_yrs') +
               nodematch('agency_lvl',diff=TRUE) +
               edgecov(TCdist,attr='distance') +
               edgecov(TCcnt,attr='contact'),
               control=control.ergm(seed=40))
summary(DSmod3)

##
## =========================
## Summary of model fit
## =========================
##
## Formula:   TCdiss ~ edges +
##            nodecov("tob_yrs") +
##            nodematch("agency_lvl",diff = TRUE) +
##            edgecov(TCdist, attr = "distance") +
##            edgecov(TCcnt,  attr = "contact")
##
```

```
## Iterations:   5 out of 20
##
## Monte Carlo MLE Results:
##                            Estimate Std. Error
## edges                     -4.850619   0.629666
## nodecov.tob_yrs            0.128750   0.028270
## nodematch.agency_lvl.1     1.795102   0.625698
## nodematch.agency_lvl.2    -0.646015   0.508164
## nodematch.agency_lvl.3     1.721850   0.546870
## edgecov.distance          -0.000184   0.000253
## edgecov.contact            1.124253   0.146957
##                           MCMC % p-value
## edges                         0  <1e-04
## nodecov.tob_yrs               0  <1e-04
## nodematch.agency_lvl.1        0   0.0044
## nodematch.agency_lvl.2        0   0.2046
## nodematch.agency_lvl.3        0   0.0018
## edgecov.distance              0   0.4683
## edgecov.contact               0  <1e-04
##
##        Null Deviance: 416  on 300  degrees of freedom
##   Residual Deviance: 237  on 293  degrees of freedom
##
## AIC: 251    BIC: 277    (Smaller is better.)
```

The results show that frequency of contact is positively associated with dissemination, as we hypothesized. On the other hand, physical distance between agencies does not appear to be associated with dissemination.

11.2.5 Including Local Structural Predictors (Dyad Dependency)

The final type of predictor that can be included in ERGMs is information about local structural properties. Remember that we want to model how an entire network looks and operates. By including some information about local structural tendencies (such as the tendency for directed ties to be reciprocated), we can often produce models that are much better fits to the observed, empirical network. These types of predictors lead to what are called dyadic-dependency models, and these present many more computational and statistical challenges (Harris 2013). From the applied analyst perspective, when using these types of predictors you should expect that the execution time may go up dramatically, and that you are much more likely to run into problems of model degeneracy and convergence failure.

In recent years, new types of local structural predictor terms have been discovered and developed that at least partially avoid the most problematic model stability

and convergence issues (Snijders and Pattison 2006). The three most commonly
used of these terms are GWDegree (geometrically weighted degree distribution),
GWESP (geometrically weighted edgewise shared partnerships), and GWDSP (geometrically weighted dyadwise shared partnerships). See Harris' monograph (2013)
for more details on how to choose and interpret these local structure predictors.

For our example, we include GWESP. This measures the effect of local clustering
(or transitivity) on the likelihood of observing a dissemination tie. The model results
do indicate that there is transitivity in the Indiana network, and that it is positively
associated with dissemination.

```
DSmod4 <- ergm(TCdiss ~ edges +
            nodecov('tob_yrs') +
            nodematch('agency_lvl',diff=TRUE) +
            edgecov(TCdist,attr='distance') +
            edgecov(TCcnt,attr="contact") +
            gwesp(0.7, fixed=TRUE),
            control=control.ergm(seed=40))

## Starting maximum likelihood estimation via MCMLE:
## Iteration 1 of at most 20:
## The log-likelihood improved by 0.5168
## Step length converged once.
## Increasing MCMC sample size.
## Iteration 2 of at most 20:
## The log-likelihood improved by 0.03238
## Step length converged twice. Stopping.
##
## This model was fit using MCMC.
## To examine model diagnostics and check
## for degeneracy, use the mcmc.diagnostics()
## function.

summary(DSmod4)

##
## ==========================
## Summary of model fit
## ==========================
##
## Formula:    TCdiss ~ edges + nodecov("tob_yrs") +
##             nodematch("agency_lvl", diff = TRUE) +
##             edgecov(TCdist, attr = "distance") +
##             edgecov(TCcnt,attr = "contact") +
##             gwesp(0.7, fixed = TRUE)
##
## Iterations:  2 out of 20
```

```
##
## Monte Carlo MLE Results:
##                              Estimate Std. Error
## edges                       -6.326172    0.960886
## nodecov.tob_yrs              0.097673    0.029612
## nodematch.agency_lvl.1       1.557478    0.595676
## nodematch.agency_lvl.2      -0.266426    0.471595
## nodematch.agency_lvl.3       1.506054    0.526105
## edgecov.distance            -0.000154    0.000243
## edgecov.contact              1.039252    0.145160
## gwesp.fixed.0.7              0.877169    0.377682
##                              MCMC % p-value
## edges                             0   <1e-04
## nodecov.tob_yrs                   0   0.0011
## nodematch.agency_lvl.1            0   0.0094
## nodematch.agency_lvl.2            0   0.5725
## nodematch.agency_lvl.3            0   0.0045
## edgecov.distance                  0   0.5279
## edgecov.contact                   0   <1e-04
## gwesp.fixed.0.7                   0   0.0209
##
##       Null Deviance: 416  on 300  degrees of freedom
##   Residual Deviance: 231  on 292  degrees of freedom
##
## AIC: 247    BIC: 276    (Smaller is better.)
```

11.3 Examining Exponential Random Graph Models

11.3.1 Model Interpretation

The fitted ERGM objects contain a lot of information about the parameter estimates, simulated networks, and model fit. The most important information is included in the model summary output. Pay particular attention to any messages about convergence issues, as they typically indicate estimation problems that should be addressed.

The parameter estimates themselves, along with their standard errors, can be interpreted similarly to logistic regression parameters. The p-values are associated with the ratio of the parameter estimates to their standard errors, which are distributed as Wald test statistics. Although not presented in the default summary output, the individual parameter estimates can be exponentiated to produce odds-ratios.

The AIC and BIC values are related to overall model fit, where lower numbers indicate better fit of the model to the observed network. Note how the AIC and BIC values were getting smaller as we progressively added predictors to the dissemination model. This is telling us that we were adding useful predictors to our ERGM.

Finally, as with any type of multivariate statistical model, it is often hard to understand the model by examining the individual parameter estimates. It is usually helpful, then, to use the fitted model to produce sets of forecasts that can be plotted or examined to see how the model works across different profiles or ranges of predictor values.

```
prd_prob1 <- plogis(-6.31 + 2*1*.099 + 1.52 +
                    4*1.042 + .858*(.50^4))
prd_prob1

  ## [1] 0.408

prd_prob2 <- plogis(-6.31 + 2*5*.099 +
                    1*1.042 + .858*(.50^4))
prd_prob2

  ## [1] 0.0144
```

As a simple example, based on our final model we predict the likelihood of a dissemination tie being observed between two agencies, where they both have been working in tobacco control for 1 year, they both are national-level agencies, they have weekly contact, and the network has an average level of transitivity. We ignore the effects of distance, given the small parameter value and lack of significance. For that predictor profile, the probability of observing a dissemination tie is 41 %. In comparison, for two agencies that have been working in tobacco control for 5 years, that are at different levels (local and national, for example), and only have contact yearly, the estimated probability is just 1.4 %. (See Harris 2013, for more in-depth worked examples, as well as details on how the forecasting handles the local structural parameters such as GWESP here.)

11.3.2 Model Fit

The ergm package includes a number of tools that can be used to examine the fit of the network model to the data. First, make sure that there were no major problems with convergence, and that the parameter values, standard errors, and p-values make sense. AIC values can also be useful, especially examining how AIC (and BIC) is reduced as more predictors are added to the model.

The simulations underlying the MCMC algorithms also provide useful information for judging model fit. The following procedure compares selected network properties of the simulated networks based on our final model to those same network characteristics of the observed Indiana tobacco control network. Specifically, here we will examine the geodesic distances, the distribution of edgewise shared partners, the degree distribution, and the triad census (frequency of different patterns of triangles).

```
DSmod.fit <- gof(DSmod4,
                 GOF = ~distance + espartners +
                 degree + triadcensus,
                 burnin=1e+5, interval = 1e+5)
summary(DSmod.fit)
```

```
##
## Goodness-of-fit for minimum geodesic distance
##
##       obs min    mean max MC p-value
## 1     103  88 102.90 116       1.00
## 2     197 123 169.44 200       0.04
## 3       0   0   8.90  34       0.20
## 4       0   0   0.02   1       1.00
## Inf     0   0  18.74  69       0.80
##
## Goodness-of-fit for edgewise shared partner
##
##         obs min   mean max MC p-value
## esp0      1   0   0.76   3       1.00
## esp1      8   0   4.94  14       0.28
## esp2     10   3  10.28  21       1.00
## esp3      7   7  17.68  34       0.02
## esp4     15   8  20.21  32       0.24
## esp5     11   7  16.77  25       0.32
## esp6      9   4  12.83  22       0.48
## esp7     14   3   7.73  16       0.12
## esp8     14   0   5.08  13       0.00
## esp9      5   0   2.77   8       0.32
## esp10     4   0   1.74   7       0.28
## esp11     1   0   1.19   3       1.00
## esp12     1   0   0.61   4       0.88
## esp13     2   0   0.13   1       0.00
## esp14     1   0   0.17   2       0.32
## esp15     0   0   0.01   1       1.00
##
## Goodness-of-fit for degree
##
##     obs min mean max MC p-value
## 0     0   0 0.79   3       0.80
## 1     1   0 0.61   3       0.96
## 2     2   0 1.07   4       0.56
## 3     3   0 1.02   4       0.14
## 4     2   0 1.22   4       0.66
## 5     1   0 1.70   6       0.98
```

```
## 6      2     0  1.87    5        1.00
## 7      2     0  2.47    7        1.00
## 8      0     0  2.76    7        0.14
## 9      1     0  2.57    8        0.54
## 10     3     0  2.25    6        0.84
## 11     2     0  2.15    5        1.00
## 12     1     0  1.45    5        1.00
## 13     1     0  1.05    4        1.00
## 14     2     0  0.62    3        0.20
## 15     1     0  0.22    2        0.40
## 16     0     0  0.13    1        1.00
## 17     0     0  0.05    1        1.00
## 18     0     0  0.08    1        1.00
## 19     0     0  0.11    1        1.00
## 20     0     0  0.24    1        1.00
## 21     0     0  0.22    1        1.00
## 22     0     0  0.24    1        1.00
## 23     0     0  0.09    1        1.00
## 24     1     0  0.02    1        0.04
##
## Goodness-of-fit for triad census
##
##     obs min mean max MC p-value
## 0  832 616  756 911       0.20
## 1  759 787  881 944       0.00
## 2  517 407  502 625       0.72
## 3  192 114  161 221       0.18
```

The goodness-of-fit object stores the results of those comparisons. If the model is doing an adequate or good job of describing the observed network, then we would expect to see that the simulated networks look like the observed network. For each possible value of the selected network statistic, the frequency of the observed value is reported alongside the minimum, mean, and maximum values across 100 (by default) randomly simulated networks. The Monte Carlo empirical p-values are also reported and these are the proportion of the simulated values of the statistic that are at least as extreme as the observed value. Thus, small p-values (traditionally <0.05) indicate cases where the model is not able to produce the particular network characteristic (i.e., poor fit).

Examination of DSmod.fit suggests that our relatively simple model with seven predictors is doing a fairly good job of capturing the structural patterns in TCdiss. Out of 50 network statistics, only four show poor fit.

In addition to examining the actual goodness-of-fit object, ergm can also easily produce informative plots of the goodness-of-fit. The resulting plot produces one panel for each of the four network statistics. Each panel includes box-plots and 95 % empirical confidence intervals (light grey lines) that show the variability of the individual network statistic across the simulated networks. The thick black line indicates the value of the same statistic for the observed network. A good-fitting model will thus have the black line sitting inside the confidence-range bands. In Fig. 11.3 we can see that our final model does a generally good job, but we can also see that our model tends to produce networks that underestimate the number of dyads with geodesics of two, while overestimating the dyads with geodesics of three.

```
op <- par(mfrow=c(2,2))
plot(DSmod.fit,cex.axis=1.6,cex.label=1.6)
par(op)
```

11.3.3 Model Diagnostics

More detailed diagnostic information about the model estimation can be produced with a call to the mcmc.diagnostics() function. This is useful for seeing how the MCMC estimation process is running 'under the hood,' and is particularly important if you run into convergence problems. The diagnostics report includes statistical details for all of the covariates in the model, and also reports information on how the model behaves over time. The plots produced display the MCMC chain over time, and a resulting histogram. Both types of plots should show estimates that are centered around 0 (Because of its length, only the diagnostics plots are shown here.) (Fig. 11.4).

```
mcmc.diagnostics(DSmod4)
```

11.3.4 Simulating Networks Based on Fit Model

The fitted ERGM can be used to produce one or more simulated networks that can then be examined or analyzed as if it were an observed network. For example, this is what a simulated network based on our final model looks like compared to the observed network. Note that the simulated network will have the same node attributes as the empirical network (Fig. 11.5).

```
sim4 <- simulate(DSmod4, nsim=1, seed=569)
summary(sim4,print.adj=FALSE)

  ## Network attributes:
  ##    vertices = 25
  ##    directed = FALSE
```

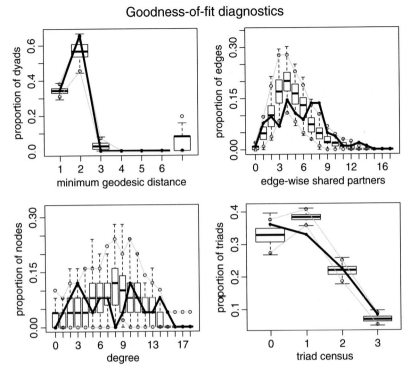

Fig. 11.3 Goodness-of-fit plots for final tobacco control model

```
##     hyper = FALSE
##     loops = FALSE
##     multiple = FALSE
##     bipartite = FALSE
##     title = IN_Diffusion
##  total edges = 115
##     missing edges = 0
##     non-missing edges = 115
##  density = 0.383
##
## Vertex attributes:
##
##  agency_cat:
##     numeric valued attribute
##     attribute summary:
##     Min. 1st Qu.  Median   Mean 3rd Qu.    Max.
##     1.00    2.00    2.00   3.24    5.00    6.00
##
```

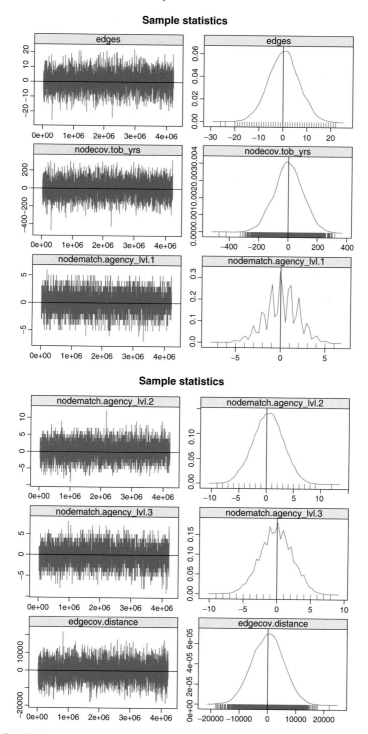

Fig. 11.4 MCMC diagnostics (partial)

```
##   agency_lvl:
##     numeric valued attribute
##     attribute summary:
##     Min. 1st Qu.  Median   Mean 3rd Qu.    Max.
##     1.00    1.00    2.00   2.04    3.00    3.00
##
##   lead_agency:
##     numeric valued attribute
##     attribute summary:
##     Min. 1st Qu.  Median   Mean 3rd Qu.    Max.
##     0.00    0.00    0.00   0.04    0.00    1.00
##
##   tob_yrs:
##     numeric valued attribute
##     attribute summary:
##     Min. 1st Qu.  Median   Mean 3rd Qu.    Max.
##     1.00    3.00    4.50   6.76    9.00   21.00
##   vertex.names:
##     character valued attribute
##     25 valid vertex names
##
## No edge attributes
```

```
op <- par(mfrow=c(1,2),mar=c(0,0,2,0))
lvlobs <- TCdiss %v% 'agency_lvl'
plot(TCdiss,usearrows=FALSE,
     vertex.col=lvl+1,
     edge.lwd=0.5,edge.col="grey75",
     main="Observed TC network")
lvl4 <- sim4 %v% 'agency_lvl'
plot(sim4,usearrows=FALSE,
     vertex.col=lvl4+1,
     edge.lwd=0.5,edge.col="grey75",
     main="Simulated network - Model 4")
par(op)
```

Observed TC network   **Simulated network - Model 4**

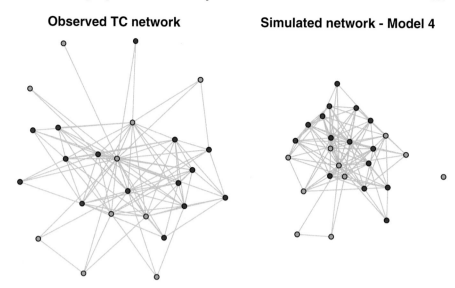

Fig. 11.5 Comparison of simulated network to observed tobacco control network

Chapter 12
Dynamic Network Models

*What came first–the music or the misery? Did I listen to the music because I was miserable? Or was I miserable because I listened to the music? Do all those records turn you into a melancholy person? (*Nick Hornby, *High Fidelity.)*

12.1 Introduction

Exponential random graph models, as presented in Chap. 11, allow for sophisticated and powerful modeling of network structures and relationships. Generative models of networks can be built using a wide variety of predictors, including node characteristics, dyad characteristics, local structural characteristics, and even other network relations. Substantive hypotheses can be tested with ERGM models, and estimated models can be explored with the rich simulation and goodness-of-fit tools that are provided by the `ergm` package.

However, ERGMs are generally limited to cross-sectional network data. Social networks, by their very nature, are dynamic. In particular, network ties are formed, maintained, and sometimes dissolved over time. These dynamic ties processes may be driven by a number of social processes, including characteristics of the actors, dyads, and local network structures. For example, one student may become a friend with another student partly because of the characteristics of the alter (e.g., attractiveness), partly because of their own similarity on some behavioral characteristic (e.g., they both like the same type of music), or because of other local network structures (e.g., they both are already friends with the same other student). This chapter covers *stochastic actor-based models for network dynamics* that are included in the `RSiena` package, and which can be used to build models and test hypotheses about networks as they change over time.

12.1.1 Dynamic Networks

Networks can change over time in two fundamental ways. First, networks can grow or shrink over time, leading to changes in network composition. Second, as suggested above, network ties can change among the network members. The modeling methods discussed in this chapter apply primarily to changes of the second type.

© Springer International Publishing Switzerland 2015
D.A. Luke, *A User's Guide to Network Analysis in R*, Use R!,
DOI 10.1007/978 3 319 23883-8_12

(Although RSiena can handle networks that have some changes in network composition, those changes themselves are not modeled.)

One of the challenges and opportunities for modeling network dynamics is that overall network characteristics can often be the result of multiple underlying social mechanisms. For example, homophily is the tendency for people (or other social entities) to associate with others who are similar to them. Social networks tend to exhibit strong homophily, and this pattern has been observed for decades across many areas of social and health sciences. There are at least two social mechanisms that can account for homophily – social selection and social influence. Social selection occurs when an actor selects or forms a new social tie with another actor who is similar to her on some relevant characteristic. Social influence, on the other hand, acts across existing social ties. Social influence occurs when the behavior of one actor is changed to become more similar (or dissimilar) to the behavior of one or more other actors.

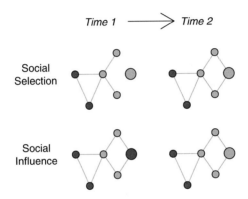

Fig. 12.1 Comparison of social selection to social influence

Figure 12.1 visually depicts these two mechanisms. The colors designate some node characteristic, such as smoking status. Blue nodes are smokers and green nodes are non-smokers, for example. The top row illustrates social selection – the focal actor (large blue node) starts out unconnected from the network. By time 2, the actor has formed two new ties with other actors who are the same as him regarding smoking status. The second row shows how social influence operates. Here, the focal actor starts out connected to the network, but differs from his friends on smoking status. At time 2 he has changed his behavior to match those of his friends. Note that the time 2 networks are identical, and show strong homophily. However, we see that this homophily can arise from two quite different mechanisms. Distinguishing these mechanisms with real network data has important scientific and applied consequences. For example, public health policies could use word-of-mouth communication strategies to disseminate prevention messages across social networks.

This policy would likely have greater effect if social influence mechanisms are the primary dynamic, where social ties are already in place and messages are more likely to pass from person to person. In contrast, if social selection is the primary mechanism, then a word-of-mouth campaign may have less influence where individuals are choosing their new relationships based on behavioral similarity.

Disentangling the effects of social influence and social selection is not possible to do with most network modeling techniques such as ERGM. This is mainly because ERGMs are generally limited to cross-sectional network data. Accurately assessing dynamic network mechanisms that operate over time requires dynamic modeling techniques such as those in RSiena.

12.1.2 RSiena

SIENA stands for *Simulation Investigation for Empirical Network Analysis*, and is a set of analytic tools that can be used to model longitudinal network data, according to the stochastic actor-oriented model (SAOM) of Snijders and his colleagues (Snijders et al. 2010). RSiena is the R package that contains the SIENA model estimation functions, as well as a wide variety of supporting tools to plot, diagnose, and examine the estimated models and simulated networks. The core SAOM is a type of actor or agent-based simulation model – it uses estimation techniques following a Markov process that assumes that future changes in network states (i.e., formation or dissolution of a tie) are based probabilistically on the current state of the complete network (Snijders et al. 2010).

RSiena combines the power of stochastic network modeling with longitudinal analysis. This opens up many analytic possibilities. RSiena can be used to model the evolution of one-mode networks, two-mode networks (see Chap. 9), and the co-evolution of one-mode or two- mode networks with behavior. This last type of model is what allows examination of social influence and social selection processes in the joint evolution of friendship and smoking, for example.

With this power comes a fair amount of complexity. In particular, handling missing data, analyzing longitudinal network data when there are changes in composition (i.e., nodes that enter or leave over time), selecting from hundreds of potential parameter effects to include in the model, and dealing with estimation convergence issues are all challenges that are beyond the ability of this short chapter to handle in any detail. Instead, this chapter presents an introduction to RSiena analysis, using an example longitudinal network dataset that has been constructed to help illustrate a basic approach to dynamic network modeling. This should help readers get started using RSiena, but it is important to consult the many excellent papers, tutorials, and documentation that are available. The manual for RSiena is a good place to start, it is available at http://www.stats.ox.ac.uk/~snijders/siena/siena_r.htm.

12.2 Data Preparation

`RSiena` requires longitudinal (or panel) network data that have been collected from the same network at two, or preferably more, timepoints. The `Coevolve` dataset that is included in `UserNetR` has been developed to support exploration of a simple co-evolution model. The Coevolve data are in the form of a list of four `igraph` networks. These are friendship networks among 37 students, measured at four points in time (or waves). These data are based on an actual school-based directed friendship network presented in Valente (2010). The original network was cross-sectional. Here, we have added three additional fictional waves of data that show changes both in tie formation as well as changes in a fictional smoking status variable.

In constructing the Coevolve networks, the following informal change rules were used to create the new waves:

1. At each wave one smoker was randomly changed to non-smoker and three non-smokers were changed to smokers, for a net gain of two smokers. The new smokers were more likely to be network members who were connected to other smokers.
2. At each wave, 10 % of the existing ties were randomly deleted. Then, the same number of new ties were formed. This maintains the same overall density over time.
3. When adding new directed ties, the following rules were used:

    (a) Pick somebody who has the same smoking status
    (b) Pick somebody who is popular (i.e., high indegree)
    (c) Reciprocate an existing tie

These informal rules were used to build in dynamics that are somewhat realistic, but also simple enough to detect with a basic RSiena model. To examine the networks, they can be plotted using `igraph`. The data are stored as a list object, so they should be extracted first.

```
library(igraph)
library("UserNetR")
data(Coevolve)
fr_w1 <- Coevolve$fr_w1
fr_w2 <- Coevolve$fr_w2
fr_w3 <- Coevolve$fr_w3
fr_w4 <- Coevolve$fr_w4
```

Figure 12.2 shows the four waves of friendship data. Node shape conveys gender (circle = female; square = male) and smoking status is conveyed by node color (green = non-smoker; blue = smoker). The increase in smoking status over time is fairly evident, and smoking status does seem to become more clustered, at least for males.

```
colors <- c("darkgreen","SkyBlue2")
shapes <- c("circle","square")
coord <- layout.kamada.kawai(fr_w1)
op <- par(mfrow=c(2,2),mar=c(1,1,2,1))
plot(fr_w1,vertex.color=colors[V(fr_w1)$smoke+1],
    vertex.shape=shapes[V(fr_w1)$gender],
    vertex.size=10,main="Wave 1",vertex.label=NA,
    edge.arrow.size=0.5,layout=coord)
plot(fr_w2,vertex.color=colors[V(fr_w2)$smoke+1],
    vertex.shape=shapes[V(fr_w2)$gender],
    vertex.size=10,main="Wave 2",vertex.label=NA,
    edge.arrow.size=0.5,layout=coord)
plot(fr_w3,vertex.color=colors[V(fr_w3)$smoke+1],
    vertex.shape=shapes[V(fr_w3)$gender],
    vertex.size=10,main="Wave 3",vertex.label=NA,
    edge.arrow.size=0.5,layout=coord)
plot(fr_w4,vertex.color=colors[V(fr_w4)$smoke+1],
    vertex.shape=shapes[V(fr_w4)$gender],
    vertex.size=10,main="Wave 4",vertex.label=NA,
    edge.arrow.size=0.5,layout=coord)
par(op)
```

Table 12.1 presents some basic descriptive statistics of the Coevolve network data. As can be seen, the size and density of the networks remain constant, but the number of smokers increases over time, as does the modularity based on smoking status. Typically, detailed examination of network graphics and descriptive statistics would happen prior to jumping into the dynamic modeling.

Wave	Size	Density	Avg.InDegree	Smokers	Modularity
Wave 1	37	0.134	4.838	8	0.001
Wave 2	37	0.134	4.838	10	0.044
Wave 3	37	0.134	4.838	12	0.077
Wave 4	37	0.134	4.838	14	0.129

Table 12.1 Characteristics of Coevolve networks across four waves

The plots and descriptive statistics suggest that there are some network and behavioral dynamics that can be modeled using RSiena. In the rest of this chapter we will explore a simple co-evolution model of friendship ties and smoking behavior. The outline of the model building process has three main steps: (1) data preparation; (2) model estimation; and (3) model exploration and testing.

Before starting, make sure to download and install the RSiena package. Like most R packages, it is available through the CRAN repository. However, the RSiena developers make newer versions of the package available at their website: http://www.stats.ox.ac.uk/~snijders/siena. This version, called

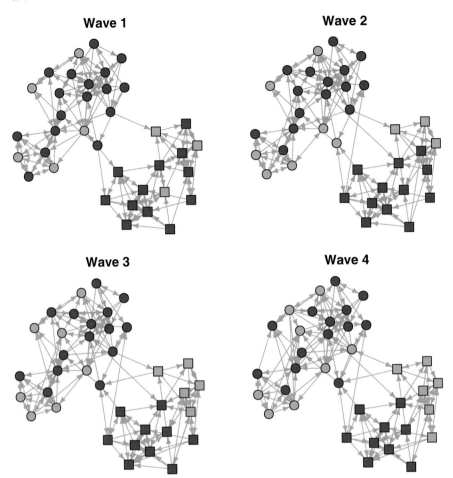

Fig. 12.2 Changes over time in Coevolve networks

RSienaTest, is the most up to date version and is used here. (For example, at the time of writing this chapter RSiena from CRAN was version 1.1-232, while RSienaTest was version 1.1-284.)

RSiena does not use the traditional method of specifying a statistical formula for a modeling function, the way that ergm, lm, or in fact most other R statistical modeling procedures do. Instead, the modeling function (called siena07) is applied to a set of RSiena objects, which must minimally include a *data* object (containing all network and covariate data), an *effects* object (containing all of the parameter effects to be included in the model), and an *algorithm* object (which controls most of the modeling options).

The first step, then, is packaging the network data in a way that RSiena can understand. RSiena can handle six different types of variables. A *network* variable is the basic dependent variable in an RSiena model, and can be a one-mode or two-mode network. A *behavior* variable is another type of dependent variable. It is a node characteristic that changes over time, the evolution of which may considered in a co-evolutionary model. In our example dataset, smoking status will be handled as a behavior variable. Then there are four types of variables that are all handled as covariates. A *coCovar* is a constant node attribute that does not change over time (e.g., gender). A *varCovar*, on the other hand, is an attribute that does change over time. (Note that a *behavior* variable is a type of varying covariate, but one that is being treated as a dependent variable.) Similar to `ergm`, `RSiena` can also handle dyadic covariates. A *coDyadCovar* is a constant dyadic covariate (for example, a kinship relationship), while *varDyadCovar* is a dyadic covariate that changes over time.

For our Coevolve data, we have one dependent network variable (the friendship ties), one constant covariate (gender), and one varying covariate that will be handled as a *behavior* dependent variable (smoking status). `RSiena` cannot handle `igraph` or `statnet` data directly, it expects data in the form of raw arrays, matrices, or vectors. So, some of the data management approaches covered in Chap. 3 will be useful here. The first step is to transform the data into raw sociomatrices.

```
library(RSienaTest)
matw1 <- as.matrix(get.adjacency(fr_w1))
matw2 <- as.matrix(get.adjacency(fr_w2))
matw3 <- as.matrix(get.adjacency(fr_w3))
matw4 <- as.matrix(get.adjacency(fr_w4))
matw1[1:8,1:8]
```

```
##         [,1] [,2] [,3] [,4] [,5] [,6] [,7] [,8]
## [1,]      0    0    0    0    0    0    1    1
## [2,]      0    0    0    0    0    0    1    0
## [3,]      0    1    0    0    0    0    0    0
## [4,]      0    0    0    0    0    0    1    0
## [5,]      0    0    0    0    0    0    0    0
## [6,]      0    0    0    0    0    0    1    0
## [7,]      1    0    0    0    0    1    0    1
## [8,]      1    0    0    0    0    0    1    0
```

Then, a dependent variable object is created with `sienaDependent`. This expects a stacked array with each array corresponding to one of the network waves. By default, `sienaDependent` expects data in the form of a sparse matrix (from the `Matrix` package). Here, the data are simple full matrices, so the `sparse` option must be set to false.

```
fr4wav<-sienaDependent(array(c(matw1,matw2,matw3,matw4),
                       dim=c(37,37,4)),sparse=FALSE)
class(fr4wav)
```

```
## [1] "sienaDependent"
```

```
fr4wav
```

```
## Type           oneMode
## Observations 4
## Nodeset        Actors (37 elements)
```

The only problem with this approach is that these sociomatrices will get quite
large for larger networks. As discussed in Chap. 3, edgelists are preferable for handling large networks. RSiena cannot natively handle edgelists, so they must be
transformed into sparse matrices as implemented by the Matrix package. The
following code produces the same friendship RSiena dependent variable, but via
edgelists instead of sociomatrices.

```
library(Matrix)
w1 <- cbind(get.edgelist(fr_w1), 1)
w2 <- cbind(get.edgelist(fr_w2), 1)
w3 <- cbind(get.edgelist(fr_w3), 1)
w4 <- cbind(get.edgelist(fr_w4), 1)
w1s <- spMatrix(37, 37, w1[,1], w1[,2], w1[,3])
w2s <- spMatrix(37, 37, w2[,1], w2[,2], w2[,3])
w3s <- spMatrix(37, 37, w3[,1], w3[,2], w3[,3])
w4s <- spMatrix(37, 37, w4[,1], w4[,2], w4[,3])
fr4wav2 <- sienaDependent(list(w1s,w2s,w3s,w4s))
fr4wav2
```

```
## Type           oneMode
## Observations 4
## Nodeset        Actors (37 elements)
```

Once the RSienda dependent variable is constructed, then other data objects
such as covariates can be created. Gender is stored in the igraph networks as
a vertex characteristic, so it is easy to extract that to create the coCoVar object.
Gender is coded 1 for female and 2 for males. The default for RSiena is to center
any covariate. This does not make much sense here, so it is turned off.

```
gender_vect <- V(fr_w1)$gender
table(gender_vect)
```

```
## gender_vect
##  1  2
## 22 15
```

```
gender <- coCovar(gender_vect,centered=FALSE)
gender
```

```
## [1] 1 1 1 1 2 1 1 1 1 2 1 2 1 2 2 2 1 1 1 1 1 2 1
## [24] 2 2 2 1 2 2 1 2 1 1 1 1 2 2
```

```
## attr(,"class")
## [1] "coCovar"
## attr(,"centered")
## [1] FALSE
## attr(,"nodeSet")
## [1] "Actors"
```

Smoking status is our behavior variable for the co-evolution model. RSiena expects an N×W matrix, with N (actor) rows and W (wave) columns. Again, we extract this information from the igraph objects. Because a behavior variable is a type of dependent variable, the sienaDependent function is used, but we specify that this is a *behavior* variable.

```
smoke <- array(c(V(fr_w1)$smoke,V(fr_w2)$smoke,
           V(fr_w3)$smoke,V(fr_w4)$smoke),dim=c(37,4))
smokebeh <- sienaDependent(smoke,type = "behavior")
smokebeh
```

```
## Type          behavior
## Observations  4
## Nodeset       Actors (37 elements)
```

Finally, all the individual variable objects are packaged together into a single RSiena data object.

```
friend <- sienaDataCreate(fr4wav,smokebeh,gender)
friend
```

```
## Dependent variables:  fr4wav, smokebeh
## Number of observations: 4
##
## Nodeset                  Actors
## Number of nodes              37
##
## Dependent variable fr4wav
## Type               oneMode
## Observations       4
## Nodeset            Actors
## Densities          0.13 0.13 0.13 0.13
##
## Dependent variable smokebeh
## Type               behavior
## Observations       4
## Nodeset            Actors
## Range              0 - 1
##
## Constant covariates:  gender
```

Once this object is created then a basic descriptive report can be generated that provides important information to examine prior to modeling.

```
print01Report(friend,modelname = 'Coevolve Example' )
```

The results of this command are not directed to the console or saved in an R object. Instead an external text file is created in the working directory, named in this case 'Coevolve Example.out'. It can be viewed using any text editor such as Notepad. The report contains a variety of information about the RSiena data, including information about any missing data, the degree distribution pattern observed across the waves of network data, and summary information about each dependent variable and covariate. A critical piece of information is found near the bottom of the report under the heading 'Change in Networks.' The Jaccard index, which is a measure of similarity, is calculated on the tie variables for each consecutive pair of waves. Although there needs to be enough change between the observation periods to allow for modeling, too much change would imply that the assumption of gradual change is not tenable. The authors of RSiena suggest that Jaccard values should be higher than 0.3 (Snijders et al. 2010). For the Coevolve data, we see Jaccard values of greater than 0.8, suggesting a fairly high level of stability over time.

12.3 Model Specification and Estimation

12.3.1 Specification of Model Effects

Once the RSiena data have been put into the correct format, model specification and building can proceed. In Chap. 11 we saw that stochastic network models can have a wide variety of parameters that test hypotheses about node attributes, similarity of node attributes between dyads, tie attributes, and local network structural properties. Longitudinal network models allow for an even larger set of potential parameters, and choosing the theoretically appropriate set of parameters can be challenging. In this chapter we will explore only a very small set of parameters. For more detailed guidance, the RSiena documentation should be read closely (especially Chap. 5 – Model specification).

The first step for model specification is to create an effects specification object with a minimal set of parameters.

```
frndeff <- getEffects( friend )
frndeff

  ##    name      effectName
  ## 1 fr4wav    constant fr4wav rate (period 1)
  ## 2 fr4wav    constant fr4wav rate (period 2)
  ## 3 fr4wav    constant fr4wav rate (period 3)
```

```
## 4 fr4wav    outdegree (density)
## 5 fr4wav    reciprocity
## 6 smokebeh rate smokebeh (period 1)
## 7 smokebeh rate smokebeh (period 2)
## 8 smokebeh rate smokebeh (period 3)
## 9 smokebeh behavior smokebeh linear shape
##    include fix    test  initialValue parm
## 1 TRUE     FALSE FALSE      2.004     0
## 2 TRUE     FALSE FALSE      2.004     0
## 3 TRUE     FALSE FALSE      2.004     0
## 4 TRUE     FALSE FALSE     -0.807     0
## 5 TRUE     FALSE FALSE      0.000     0
## 6 TRUE     FALSE FALSE      0.208     0
## 7 TRUE     FALSE FALSE      0.208     0
## 8 TRUE     FALSE FALSE      0.208     0
## 9 TRUE     FALSE FALSE      0.562     0
```

This is a basic model that only includes a small number of default effects, notably the outdegree and reciprocity effects. Typically we will add other effects that we are interested in testing or exploring. What are those effects and how are they specified?

All the effects that are available given the structure of the `friend` data set can be seen using the `effectsDocumentation` function. Here is one place (of many) where `RSiena` shows its non-R roots. Instead of sending the results to the R console, or creating a new R object, this function creates an HTML file that can then be opened in your browser.

```
effectsDocumentation(frndeff)
```

The effects documentation report is a type of 'effects dictionary' that provides all the information necessary for selecting appropriate model parameters. However, the report can be daunting – for this simple example dataset with four timepoints, one covariate (gender) and one behavior dependent variable (smokebeh), there are over 400 possible effects!

Table 12.2 presents a set of effects that can be tested, along with the information from the effects documentation that is used to correctly specify the effects in the model estimation function. To understand this information, please also refer to the 'frndeff.html' file that is produced by the `effectsDocumentation()` function.

Because we are exploring a co-evolutionary model, we will examine effects on the likelihood of tie formations (the `fr4wav` dependent variable), and the effects on changes in behavior (the `smokebeh` dependent variable). So, there are two broad types of effects, and they are distinguished by the 'Name' column in the effects documentation report. ('ED Name' in Table 12.2.) The actual specific effect that will be included is specified by the term in the 'shortName' column. Many of these effects will need to refer to a specific covariate or dependent variable, this is typically specified by the term in the 'interaction1' column. To help navigate this information for

Effect	Type	ED name	ED shortName	ED interaction1	ED row
1 - Gender homophily	Selection	fr4wav	sameX	gender	139
2 - Ego smoking effect	Selection	fr4wav	egoX	smokebeh	200
3 - Alter smoking effect	Selection	fr4wav	altX	smokebeh	197
4 - Smoking homophily	Selection	fr4wav	sameX	smokebeh	212
5 - Avg. alter influence	Influence	smokebeh	avSim	fr4wav	321
6 - Total alter influence	Influence	smokebeh	totSim	fr4wav	324
7 - Reciprocity	Structural	fr4wav	recip	NA	14
8 - Transitivity	Structural	fr4wav	transTrip	NA	17

Table 12.2 RSiena effects for Coevolve data

the example, Table 12.2 also includes the particular row number from the complete effects documentation report, but this will be accurate only if the data have been prepared exactly the same way as in this chapter.

Eight different hypotheses will be tested with the eight effects listed in the table. First, based on the pattern evident in Fig. 12.2, there appears to be a strong gender homophily effect, where students are much more likely to be friends with other students of the same gender. This is a type of social selection effect, we hypothesize that the likelihood of an ego forming a new friendship tie is higher with an alter who has the same gender. The RSiena term that will be used is 'sameX.' Next, two different social selection main effects are hypothesized. The first is the hypothesis that based on the ego's smoking status, he or she is more or less likely to form a friendship tie (egoX). Conversely, the likelihood of forming a friendship tie may be related to the smoking status of alters (altX). This may occur, for example, in schools that have a strong pro-smoking culture. I may want to be a friend with somebody because they smoke and smoking is 'cool,' regardless of whether I smoke or not. Note that we have no reason to assume either of these hypotheses are true, given how these data were constructed. Finally, the last social selection hypothesis is another homophily effect, but this time in reference to smoking. The hypothesis is that new friendship ties are more likely to be formed between two students who have the same smoking status. Note that this hypothesis uses the same shortName. In this case the interaction1 term is used to indicate that the homophily relationship refers to smoking (instead of gender).

The next two hypotheses are about potential social influences on behavior. They are similar to each other in that they focus on whether changes in smoking status can be explained by patterns of smoking of the ego's friends (to whom the ego is tied). The first hypothesis is that likelihood of changing behavior is related to the average similarity of smoking status across all tied alters (avSim). The second hypothesis is similar, but assumes that the influence is based on the total similarity across alters, instead of average. (Total similarity captures the effect of having a larger personal social network that influences behavior.)

The last two hypotheses are local structural effects. Here we will model the tendency for friendship ties to be reciprocated (recip) as well as the general pattern of transitivity (transTrip).

These effects are specified in an RSiena effects object. Typically, effects are added one at a time using the includeEffects() function. The following code adds the effects that correspond to the eight hypotheses just described.

```
frndeff <- getEffects( friend )
frndeff <- includeEffects(frndeff,sameX,
                interaction1="gender",name="fr4wav")

##   effectName  include fix   test   initialValue
## 1 same gender TRUE    FALSE FALSE              0
##   parm
## 1 0

frndeff <- includeEffects(frndeff,egoX,
                interaction1="smokebeh",name="fr4wav")

##   effectName  include fix   test   initialValue
## 1 smokebeh ego TRUE   FALSE FALSE              0
##   parm
## 1 0

frndeff <- includeEffects(frndeff,altX,
                interaction1="smokebeh",name="fr4wav")

##   effectName    include fix   test   initialValue
## 1 smokebeh alter TRUE   FALSE FALSE              0
##   parm
## 1 0

frndeff <- includeEffects(frndeff,sameX,
                interaction1="smokebeh",name="fr4wav")

##   effectName    include fix   test   initialValue
## 1 same smokebeh TRUE    FALSE FALSE              0
##   parm
## 1 0

frndeff <- includeEffects(frndeff,avSim,
                interaction1="fr4wav",name="smokebeh")

##   effectName                              include
## 1 behavior smokebeh average similarity TRUE
##   fix    test   initialValue parm
## 1 FALSE FALSE              0    0

frndeff <- includeEffects(frndeff,totSim,
                interaction1="fr4wav",name="smokebeh")
```

```
##   effectName                              include
## 1 behavior smokebeh total similarity TRUE
##   fix    test   initialValue parm
## 1 FALSE FALSE              0   0
```

```
frndeff <- includeEffects(frndeff,recip,transTrip,
                   name="fr4wav")
```

```
##   effectName            include fix    test
## 1 reciprocity           TRUE    FALSE FALSE
## 2 transitive triplets   TRUE    FALSE FALSE
##   initialValue parm
## 1            0   0
## 2            0   0
```

```
frndeff
```

```
##       name       effectName
## 1  fr4wav     constant fr4wav rate (period 1)
## 2  fr4wav     constant fr4wav rate (period 2)
## 3  fr4wav     constant fr4wav rate (period 3)
## 4  fr4wav     outdegree (density)
## 5  fr4wav     reciprocity
## 6  fr4wav     transitive triplets
## 7  fr4wav     same gender
## 8  fr4wav     smokebeh alter
## 9  fr4wav     smokebeh ego
## 10 fr4wav     same smokebeh
## 11 smokebeh rate smokebeh (period 1)
## 12 smokebeh rate smokebeh (period 2)
## 13 smokebeh rate smokebeh (period 3)
## 14 smokebeh behavior smokebeh linear shape
## 15 smokebeh behavior smokebeh average similarity
## 16 smokebeh behavior smokebeh total similarity
##    include fix   test   initialValue parm
## 1  TRUE    FALSE FALSE       2.004    0
## 2  TRUE    FALSE FALSE       2.004    0
## 3  TRUE    FALSE FALSE       2.004    0
## 4  TRUE    FALSE FALSE      -0.807    0
## 5  TRUE    FALSE FALSE       0.000    0
## 6  TRUE    FALSE FALSE       0.000    0
## 7  TRUE    FALSE FALSE       0.000    0
## 8  TRUE    FALSE FALSE       0.000    0
## 9  TRUE    FALSE FALSE       0.000    0
## 10 TRUE    FALSE FALSE       0.000    0
## 11 TRUE    FALSE FALSE       0.208    0
```

```
## 12 TRUE      FALSE FALSE      0.208   0
## 13 TRUE      FALSE FALSE      0.208   0
## 14 TRUE      FALSE FALSE      0.562   0
## 15 TRUE      FALSE FALSE      0.000   0
## 16 TRUE      FALSE FALSE      0.000   0
```

12.3.2 Model Estimation

The last step before estimating the model is to set up any algorithm options that are required. In this case, other than specifying a title string for the model, the algorithm object will include all default options.

```
myalgorithm <- sienaAlgorithmCreate(projname='coevolve')
```

An RSiena model is estimated using the siena07() function, and by passing the algorithm, data, and effects objects that were already created. Other options can be specified as well to control the estimation process. In the following example, the batch and verbose options are used to simplify the output. When run interactively, you may want to set these options to TRUE. The last three options are used to speed up the estimation process by using multiple cores of the computer's CPU, if available. Here, three cores, out of four, are used. Be careful about using all the CPU cores, in case other processes are being run by the operating system. See the help file for more technical details.

```
set.seed(999)
RSmod1 <- siena07( myalgorithm, data = friend,
            effects = frndeff,batch=TRUE,
            verbose=FALSE,useCluster=TRUE,
            initC=TRUE,nbrNodes=3)
```

12.4 Model Exploration

12.4.1 Model Interpretation

As is typical with any R estimation technique, the results of the modeling can be explored by viewing the contents of the model fit object. Either list the RSiena fit object directly, or use the summary() command for a more detailed output. (The formatting of the output listed here has been edited for length and legibility.)

```
summary(RSmod1)

## Estimates, standard errors and convergence t-ratios
##
## Network Dynamics
##    1. rate constant fr4wav rate (period 1)
##    2. rate constant fr4wav rate (period 2)
##    3. rate constant fr4wav rate (period 3)
##    4. eval outdegree (density)
##    5. eval reciprocity
##    6. eval transitive triplets
##    7. eval same gender
##    8. eval smokebeh alter
##    9. eval smokebeh ego
##   10. eval same smokebeh
##
##                 Estimate   Standard    Convergence
##                             Error       t-ratio
##
##    1.           1.1572  (  0.2073  )   -0.0491
##    2.           1.1410  (  0.1990  )   -0.0114
##    3.           1.1366  (  0.1948  )   -0.0273
##    4.          -2.9556  (  0.3949  )    0.0141
##    5.           0.7990  (  0.2539  )   -0.0805
##    6.           0.0860  (  0.0785  )   -0.0508
##    7.           1.1429  (  0.3174  )   -0.1147
##    8.           0.6885  (  0.3792  )   -0.0122
##    9.          -0.0847  (  0.2878  )    0.0165
##   10.           1.0975  (  0.4373  )   -0.1392
##
## Behavior Dynamics
##   11. rate rate smokebeh (period 1)
##   12. rate rate smokebeh (period 2)
##   13. rate rate smokebeh (period 3)
##   14. eval behavior smokebeh linear shape
##   15. eval behavior smokebeh average similarity
##   16. eval behavior smokebeh total similarity
##
##                 Estimate   Standard    Convergence
##                             Error       t-ratio
##
##   11.           0.3028  (  0.1684  )    0.0082
##   12.           0.3485  (  0.1963  )    0.0528
##   13.           0.3363  (  0.1949  )    0.0406
##   14.           4.0791  (  8.7065  )   -0.0218
```

```
##    15.        18.7278    ( 53.9223   )     -0.1372
##    16.        -1.4614    (  6.9254   )      0.0824
##
## Total of 2340 iteration steps.
##
```

For a coevolution model, the parameter estimates are presented in two sections. The network dynamics section contains the estimates pertaining to the tie formation (i.e., the `fr4wav` dependent variable). Conversely, the behavior dynamics section contains estimates related to changes in the network member behavior variable, here it is smoking status.

The convergence *t*-ratios are *not* traditional *t*-statistics assessing the size of the parameter estimates. Instead, they represent tests of the lack of convergence for each estimate, so small values indicate good convergence. The RSiena manual suggests that absolute values less than 0.10 indicate excellent convergence, and absolute values less than 0.15 are reasonable. Here we see that all of the network dynamics parameters have excellent convergence, while a few of the behavior parameters show only reasonable convergence.

The *rate* estimates correspond to the estimated number of opportunities for change per actor for each period (where period 1 is the time from wave 1 to wave 2). The *eval* estimates are the weights in the network evaluation function. The exact calculations for a precise interpretation of the meaning of these effects is complicated, see Snijders et al. (2010) for more details. However, they represent the relative 'attractiveness' of a particular network state for each actor. For example, the positive estimate for same gender (1.13) indicates that actors are more likely to form new ties (or maintain existing ties) with other actors who have the same gender as them.

The significance of these evaluation function weights can be determined by dividing the estimates by their standard errors. These are distributed as *t*-statistics, so any absolute values greater than 2 are significant at the 0.05 significance level.

For our example, we can see that our friendship formation is more likely with alters who have the same gender and same smoking status as the ego. Conversely, it appears that the main effects of ego smoking and alter smoking are not significant predictors of tie formation. Outdegree and reciprocity are significant structural predictors, but not transitivity. The behavior dynamics results suggest that smoking status is increasing over time (linear shape). The large positive estimate for average similarity would normally suggest that changes in smoking status are driven by the overall similarity of the smoking status for all tied alters. However, this estimate has a very large standard error, so the evaluation function weight is not being estimated with much precision. This is not too surprising, because detecting changes in actor behavior is harder than detecting changes in tie formation. There is greater power to detect tie changes because of the greater number of potential ties (on the order of the square of the size of the network for each period). Conversely, the number of behavior changes is on the order of the simple number of network members for each period. This is especially true for this example, where the network is relatively small and only a few changes in smoking status happened at each wave. So, although we know these smoking changes are 'real', the analysis does not have the required power to detect the effects.

Typically, we will make adjustments and build subsequent models based on what we learned from earlier models. For this example, we will drop a few non-significant predictors. To do this we simply update the effects object with either new predictors, or by listing the predictors that we would like to drop. A dropped predictor is indicated by the 'include = FALSE' option. Here we drop the total similarity predictor for the behavior variable as well as the transitivity predictor.

```
frndeff2 <- includeEffects(frndeff,totSim,
                           interaction1="fr4wav",
                           name="smokebeh",
                           include=FALSE)

  ## [1] effectName    include       fix
  ## [4] test          initialValue parm
  ## <0 rows> (or 0-length row.names)

frndeff2 <- includeEffects(frndeff2,transTrip,
                           name="fr4wav",
                           include=FALSE)

  ## [1] effectName    include       fix
  ## [4] test          initialValue parm
  ## <0 rows> (or 0-length row.names)

frndeff2

  ##      name      effectName
  ## 1    fr4wav    constant fr4wav rate (period 1)
  ## 2    fr4wav    constant fr4wav rate (period 2)
  ## 3    fr4wav    constant fr4wav rate (period 3)
  ## 4    fr4wav    outdegree (density)
  ## 5    fr4wav    reciprocity
  ## 6    fr4wav    same gender
  ## 7    fr4wav    smokebeh alter
  ## 8    fr4wav    smokebeh ego
  ## 9    fr4wav    same smokebeh
  ## 10   smokebeh  rate smokebeh (period 1)
  ## 11   smokebeh  rate smokebeh (period 2)
  ## 12   smokebeh  rate smokebeh (period 3)
  ## 13   smokebeh  behavior smokebeh linear shape
  ## 14   smokebeh  behavior smokebeh average similarity
  ##      include fix    test   initialValue parm
  ## 1    TRUE    FALSE FALSE      2.004      0
  ## 2    TRUE    FALSE FALSE      2.004      0
```

```
## 3   TRUE      FALSE FALSE         2.004    0
## 4   TRUE      FALSE FALSE        -0.807    0
## 5   TRUE      FALSE FALSE         0.000    0
## 6   TRUE      FALSE FALSE         0.000    0
## 7   TRUE      FALSE FALSE         0.000    0
## 8   TRUE      FALSE FALSE         0.000    0
## 9   TRUE      FALSE FALSE         0.000    0
## 10  TRUE      FALSE FALSE         0.208    0
## 11  TRUE      FALSE FALSE         0.208    0
## 12  TRUE      FALSE FALSE         0.208    0
## 13  TRUE      FALSE FALSE         0.562    0
## 14  TRUE      FALSE FALSE         0.000    0
```

```
myalgorithm <- sienaAlgorithmCreate(projname='coevol2')
```

Now the next model can be estimated. RSiena allows us to use the estimates obtained from a previous model as the starting values for the new model estimation. In this case we specify that the starting values should be based on the estimates contained in RSmod1, using the prevAns option. This is also sometimes helpful for improving the convergence of the individual weight estimates. Finally, in this second model, we use the returnDeps option. This stores some required auxiliary statistics on the simulated dependent variables for use in a subsequent exploration of goodness-of-fit.

```
set.seed(999)
RSmod2 <- siena07(myalgorithm,data = friend,
                  effects = frndeff2,
                  prevAns=RSmod1,batch=TRUE,
                  verbose=FALSE,useCluster=TRUE,
                  initC=TRUE,nbrNodes=3,
                  returnDeps=TRUE)
```

```
summary(RSmod2)
```

```
## Estimates, standard errors and convergence t-ratios
##
## Network Dynamics
##     1. rate constant fr4wav rate (period 1)
##     2. rate constant fr4wav rate (period 2)
##     3. rate constant fr4wav rate (period 3)
##     4. eval outdegree (density)
##     5. eval reciprocity
##     6. eval same gender
##     7. eval smokebeh alter
##     8. eval smokebeh ego
```

```
##     9. eval same smokebeh
##
##
##                Estimate    Standard    Convergence
##                            Error       t-ratio
##
##     1.         1.1392  (   0.223  )    -0.0454
##     2.         1.1321  (   0.213  )     0.0056
##     3.         1.1260  (   0.200  )    -0.0086
##     4.        -3.0585  (   0.422  )     0.0695
##     5.         0.8387  (   0.247  )     0.0807
##     6.         1.4113  (   0.303  )     0.0423
##     7.         0.7123  (   0.417  )     0.0127
##     8.        -0.0733  (   0.307  )     0.0721
##     9.         1.2041  (   0.486  )     0.0303
##
## Behavior Dynamics
##    10. rate rate smokebeh (period 1)
##    11. rate rate smokebeh (period 2)
##    12. rate rate smokebeh (period 3)
##    13. eval behavior smokebeh linear shape
##    14. eval behavior smokebeh average similarity
##
##
##                Estimate    Standard    Convergence
##                            Error       t-ratio
##
##    10.         0.3076  (   0.170  )     0.0508
##    11.         0.3680  (   0.249  )     0.0212
##    12.         0.3493  (   0.181  )     0.0178
##    13.         6.6261  (  37.771  )    -0.0061
##    14.        18.6088  ( 111.033  )     0.0270
##
## Total of 2140 iteration steps.
##
```

By dropping some non-significant variables and starting with previously estimated weight estimates, we have improved the convergence, now all effects have excellent convergence. The similarity effect on smoking behavior is still positive, but also still with a large standard error. A larger network, or more waves of data would likely be required to improve the standard error.

12.4.2 Goodness-of-Fit

Similar to the `ergm` package, `RSiena` includes graphical and diagnostic facilities for exploring the goodness-of-fit of an estimated model to the observed network. As in `ergm`, goodness-of-fit statistics can be calculated on properties of the simulated networks that are not formally included as predictors in the fitted models. This allows us to assess the extent to which the model can produce simulated networks that 'look like' the observed network.

Goodness-of-fit is assessed in `RSiena` using the `sienaGOF` function. A network descriptive statistic is specified and is calculated across the simulated networks at the end of each period. These values are then compared to the observed network using the Mahalanobis distance.

In the current version of the `RSienaTest` package only a small number of network statistics are built into the GOF function. One of these is the indegree distribution. In the following code, we also constrain the possible values of indegree to the range 1–10 (using the `levls` option). This matches the range of indegrees in the observed friendship network at wave 4.

```
table(degree(fr_w4,mode="in"))

  ##
  ## 1  2  3  4  5  6  7  8  9 10
  ## 2  3  7  6  5  4  6  2  1  1

gofi <- sienaGOF(RSmod2, IndegreeDistribution,
                 levls=1:10,verbose=FALSE, join=TRUE,
                 varName="fr4wav")
```

Once the GOF object is created, it can be used to plot the goodness-of-fit information. In Fig. 12.3, violin plots are used to display the variability of the descriptive statistic across the simulated networks, in this case the indegrees. The dashed grey lines represent the empirical 95 % confidence interval. The red circles are the values of the descriptive statistic for the observed network. When the circles fit inside the confidence intervals we interpret that as evidence of good fit. In this case, our second model does an excellent job of producing simulated networks that have the same or similar indegree distributions.

```
plot(gofi)
```

Although only a few statistics are currently built into `sienaGOF`, RSiena supports adding in user-supplied statistic functions. This makes it easy to assess goodness-of-fit with almost any network characteristic that is of interest. The following example (taken from the `sienaGOF-auxiliary` help file) shows how to define and then use the Holland and Leinhardt triad census (1978). The triad census gives the frequency distribution of all possible triads in a directed network. The census uses a 3-digit numeric code to designate one of the 16 possible patterns of a triad. The first digit indicates the number of reciprocated ties in the triad, the second digit is the number of oneway ties, and the third digit is the number of empty

Goodness of Fit of IndegreeDistribution

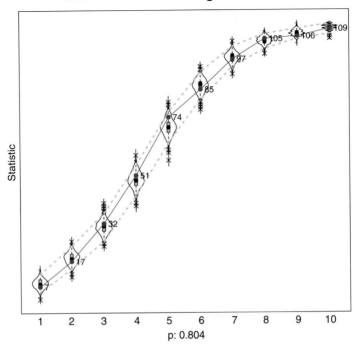

Fig. 12.3 Goodness-of-fit for indegree

ties. So, 300 is the code for a triad connected by 3 reciprocal ties, while 003 is the code for a completely unconnected triad. See Wasserman and Faust (1994) for more information.

In the following code, a new `TriadCensus` function is created. This function uses the existing `triad.census` function contained in the `statnet` package. (Note that `igraph` could also be used to calculate descriptive statistics to be used by `sienaGOF`.) This wrapper function is then used by `sienaGOF`.

```
TriadCensus <- function(i,data,sims,wave,
                    groupName,varName,levls=1:16){
  unloadNamespace("igraph") # to avoid package clashes
  require(sna)
  require(network)
  x <- networkExtraction(i,data,sims,wave,
                    groupName,varName)
  if (network.edgecount(x) <= 0){x <- symmetrize(x)}
  # because else triad.census(x) will lead to an error
  tc <- sna::triad.census(x)[1,levls]
  # names are transferred automatically
  tc
```

```
}
```

The GOF information stored in the fit object can also directly examined or summarized. The information can be plotted as before. For this type of auxiliary statistic it is advisable in the plot to center and scale.

Here we see that our second model does a pretty good job of recreating the observed pattern of directed triads. It only fails in two cases: 021C (the model overestimates the number of this type of triad with two directed ties), and 120U (the model underestimates this type of triad with one reciprocal tie and two directed ties) (Fig. 12.4).

```
goftc <- sienaGOF(RSmod2, TriadCensus,
                  varName="fr4wav",
                  verbose=FALSE, join=TRUE)
descriptives.sienaGOF(goftc)

##                003   012   102 021D 021U 021C 111D
## max          12416  6506  4301  203  273  425  454
## perc.upper   12199  6248  4055  183  247  360  431
## mean         11830  5895  3775  152  208  302  390
## median       11832  5900  3774  151  208  301  390
## perc.lower   11474  5538  3513  124  173  252  350
## min          11289  5235  3369  111  153  199  328
## obs          11771  6018  3816  137  213  232  353
##              111U  030T  030C  201 120D 120U 120C
## max           336 100.0 19.00  182 80.0 54.0 85.0
## perc.upper    301  88.0 15.00  157 70.0 46.0 69.0
## mean          262  68.3  8.11  127 57.5 35.9 53.7
## median        263  68.0  8.00  126 58.0 36.0 54.0
## perc.lower    227  51.0  3.00  101 45.0 25.0 41.0
## min           214  39.0  1.00   91 41.0 19.0 35.0
## obs           245  82.0  8.00  119 61.0 49.0 65.0
##              210   300
## max          128  67.0
## perc.upper   118  55.0
## mean         102  43.3
## median       102  43.0
## perc.lower    85  32.0
## min           74  25.0
## obs          105  36.0
```

```
plot(goftc, center=TRUE, scale=TRUE)
```

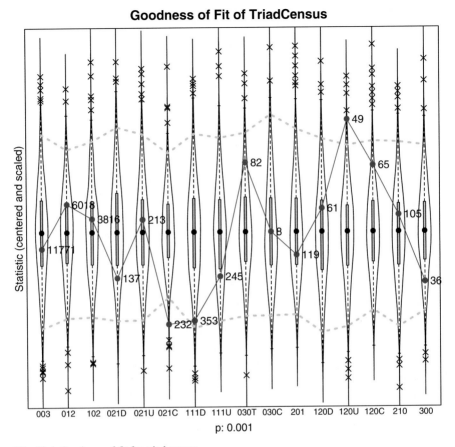

Fig. 12.4 Goodness-of-fit for triad census

12.4.3 Model Simulations

The actual simulated networks can be accessed if the `returnDeps` option has been set to true when estimating the model. (This can take a lot of memory for larger networks, so the default is false.) The simulated network information is stored in a nested list object (called `sims`) within the general fitted model object. The information is organized as an edgelist for each simulated network for each period. The default number of simulated networks is 1,000 (set by the n3 parameter in the `sienaAlgorithmCreate` function). Within a particular simulation run, one predicted network is created at the end of each period. So, in this example, with four waves of data there will be three simulated networks, one each for the ends of periods one, two, and three.

The general structure of the `sims` information can be seen with the following code. Here, the 500th simulation run is being accessed.

```
str(RSmod2$sims[[500]])
```

```
## List of 1
##  $ Data1:List of 2
##   ..$ fr4wav  :List of 3
##   .. ..$ 1: int [1:188, 1:3] 1 1 2 2 2 2 3 3 3 3...
##   .. ..$ 2: int [1:185, 1:3] 1 1 1 2 2 2 2 2 3 3...
##   .. ..$ 3: int [1:176, 1:3] 1 1 1 2 2 2 2 2 2 2...
##   ..$ smokebeh:List of 3
##   .. ..$ 1: int [1:37] 1 0 0 0 0 1 0 0 0 0 ...
##   .. ..$ 2: int [1:37] 1 1 0 0 0 1 0 0 0 0 ...
##   .. ..$ 3: int [1:37] 1 0 0 1 0 1 0 0 0 0 ...
```

The actual edgelist can be obtained as follows, again for the 500th run. The first index specifies the run number, the second index is the number of the group (RSiena can do multiple-group estimation), the third index is the number of the dependent variable, and the last index is the period number (or, equivalently, Wave - 1). For a coevolution model there are two dependent variables. The first one is the tie variable fr4wav, and the second is the behavior dependent variable smokebeh. This code, then, provides the friendship tie edgelist for the 500th run and the third period (limited to the first 25 cases).

```
RSmod2$sims[[500]][[1]][[1]][[3]][1:25,]
```

```
##         [,1] [,2] [,3]
##  [1,]    1    7    1
##  [2,]    1    8    1
##  [3,]    1   11    1
##  [4,]    2    7    1
##  [5,]    2   17    1
##  [6,]    2   21    1
##  [7,]    2   30    1
##  [8,]    2   32    1
##  [9,]    2   35    1
## [10,]    2   37    1
## [11,]    3   13    1
## [12,]    3   17    1
## [13,]    3   20    1
## [14,]    3   21    1
## [15,]    3   33    1
## [16,]    4   13    1
## [17,]    4   17    1
## [18,]    4   21    1
## [19,]    4   27    1
## [20,]    5   10    1
## [21,]    5   22    1
```

```
## [22,]    5    25    1
## [23,]    5    28    1
## [24,]    5    29    1
## [25,]    6     9    1
```

The simulated smoking status information for the same run and wave is then:

```
RSmod2$sims[[500]][[1]][[2]][[3]]
```

```
## [1] 1 0 0 1 0 1 0 0 0 0 1 0 0 1 1 1 0 0 0 0 1 0 0
## [24] 1 0 0 1 0 0 0 0 1 1 1 0 0 1
```

Using the above information, you can access and transform the data into a statnet or igraph network object. For example, the following code uses igraph to create, examine, and plot one of the simulated networks from the second RSiena model (Fig. 12.5).

```
library(igraph)
el <- RSmod2$sims[[500]][[1]][[1]][[3]]
sb <- RSmod2$sims[[500]][[1]][[2]][[3]]
fr_w4_sim <- graph.data.frame(el,directed = TRUE)
V(fr_w4_sim)$smoke <- sb
V(fr_w4_sim)$gender <- V(fr_w4)$gender
fr_w4_sim
```

```
## IGRAPH DN-- 37 176 --
## + attr: name (v/c), smoke (v/n), gender
## | (v/n), V3 (e/n)
## + edges (vertex names):
##  [1] 1 ->7   1 ->8   1 ->11  2 ->7   2 ->17  2 ->21
##  [7] 2 ->30  2 ->32  2 ->35  2 ->37  3 ->13  3 ->17
## [13] 3 ->20  3 ->21  3 ->33  4 ->13  4 ->17  4 ->21
## [19] 4 ->27  5 ->10  5 ->22  5 ->25  5 ->28  5 ->29
## [25] 6 ->9   6 ->11  6 ->15  6 ->34  7 ->1   7 ->8
## [31] 7 ->13  7 ->18  8 ->1   8 ->4   8 ->7   8 ->17
## [37] 8 ->19  8 ->21  8 ->30  8 ->32  9 ->6   9 ->7
## + ... omitted several edges
```

```
modularity(fr_w4_sim,membership = V(fr_w4_sim)$smoke+1)
```

```
## [1] 0.112
```

```
modularity(fr_w4,membership = V(fr_w4)$smoke+1)
```

```
## [1] 0.129
```

```
colors <- c("darkgreen","SkyBlue2")
coord <- layout.kamada.kawai(fr_w4)
op <- par(mfrow=c(1,2),mar=c(1,1,2,1))
plot(fr_w4,vertex.color=colors[V(fr_w4)$smoke+1],
     vertex.shape=shapes[V(fr_w4)$gender],
     vertex.size=10,main="Observed - Wave 4",
     vertex.label=NA,
     edge.arrow.size=0.5,layout=coord)
plot(fr_w4_sim,
     vertex.color=colors[V(fr_w4_sim)$smoke+1],
     vertex.shape=shapes[V(fr_w4_sim)$gender],
     vertex.size=10,main="Simulated - Wave 4",
     vertex.label=NA,
     edge.arrow.size=0.5,layout=coord)
par(op)
```

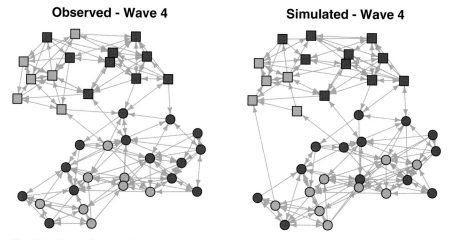

Fig. 12.5 Comparison of observed network to simulated network

Chapter 13
Simulations

*Sometimes, if you want to change a man's mind, you have to change the mind of the man next to him first. (*Megan Whalen Turner – *The King of Attolia)*

13.1 Simulations of Network Dynamics

Chapter 10 illustrated how R tools can be used to simulate networks with specific structures, often based on particular network science models. These modeled networks are useful in that they reveal social structures that may reflect reality, or are interesting for purely theoretical reasons. In any case, these are static networks.

However, social networks are dynamic. Social networks can grow or shrink over time, and their composition can similarly change. For example, friendship networks grow as people expand their friendship group, get smaller if friends move away or friendships cool. The composition of friendship networks may change dramatically during transitions (e.g., from middle school to high school or from college to work). This type of network dynamics is captured by changes in node composition (which people are in the network) and by changes in the pattern of ties connecting the nodes.

A second type of network dynamics is when some characteristic or behavior of members in a social network is influenced by the structural properties of the network itself. For example, in public health it has long been known that adolescents are more likely to start smoking if they have friends or family members in their social networks who smoke.

In the rest of this chapter two detailed examples will be presented that illustrate how R can be used to model these two broad types of social dynamics. The ability to build simulations of network dynamics reflects a particular strength of R. In particular, these network simulations are possible because of the integration of data management, statistical programming, and network analysis in R. These types of models are not possible to do in any traditional network analysis package such as Pajek, UCINet, Gephi, or NodeXL.

© Springer International Publishing Switzerland 2015
D.A. Luke, *A User's Guide to Network Analysis in R*, Use R!,
DOI 10.1007/978-3-319-23883-8_13

13.1.1 Simulating Social Selection

When network structures change over time, we assume that there is an underlying process of *social selection* that determines how new ties are formed or dissolved. (Although this process may be at least partly random, we usually want to propose a more interesting model where tie formation is driven by node characteristics, local network structure, or network history.) Social selection has been studied extensively in the social sciences, and it has been shown to partly explain *homophily*, which is the general pattern that similar types of people tend to be connected to each other (McPherson et al. 2001; also see Fig. 12.1).

In this section a dynamic model of social selection is constructed. The model is deliberately kept simple, both to make it easier to understand, but also to reflect good model building practices. In particular, it is generally a good idea when exploring computational simulations to start simple, and only add complexity once the simple model is fully understood. Also, this model is not built to emphasize efficiency (speed of execution). Efficiency could be built in later when the simulation has been fully tested and the analyst is ready to scale-up the simulation to handle larger runs.

For our model, we will assume that we have a friendship network where friendship ties can change over time, and that these changes are driven by the similarity (and dissimilarity) among network members on some abstract node characteristic. This characteristic can be thought of as a behavior, or also possibly an attitude or opinion. The characteristic is also quantitative, so that we can think of network members having more or less of this characteristic. For example, the characteristic might represent physical activity behavior, where some network members have more of this characteristic (they exercise more) or less of it. In this example, we will use the simulation model to understand how overall network homophily changes over time based on individual changes in network tie formation and dissolution.

13.1.1.1 Setting Up the Simulation

To start building and testing our simulation, we need to have a network to work with. We also want to define some basic network characteristics and model parameters that will be used as we proceed. We will start with a simple random network that has 25 members.

```
library(igraph)
N <- 25
netdum <- erdos.renyi.game(N, p=0.10)
graph.density(netdum)
mean(degree(netdum))
```

```
## [1] 0.1
## [1] 2.4
```

The network also needs to include a node characteristic that will be used to drive the network changes. This abstract behavior (Bh) can range from 0 to 1 to capture diversity from low to high. To help with interpretation later on, a categorical node characteristic (BhCat) is also calculated. The categorical variable has five levels, and can be thought of as 'very low,' 'somewhat low,' 'medium,' 'somewhat high,' and 'very high.'

```
Bh <- runif(N,0,1)
BhCat <- cut(Bh, breaks=5, labels = FALSE)
V(netdum)$Bh <- Bh
V(netdum)$BhCat <- BhCat
table(V(netdum)$BhCat)
```

```
##
## 1 2 3 4 5
## 8 5 2 4 6
```

Here is what the network looks like after setting up a color palette that maps onto the five levels of BhCat. (See Chap. 5 for details on how color palettes work.)

```
library(RColorBrewer)
my_pal <- brewer.pal(5, "PiYG")
V(netdum)$color <- my_pal[V(netdum)$BhCat]
crd_save <- layout.auto(netdum)
plot(netdum, layout = crd_save)
```

13.1.1.2 Creating an Update Function

To start the simulation building process, we will work from the inside out. That is, we start by building the inner workings that allow a network to change, and then build a simulation framework around that. The core of a dynamic network model is allowing a network to change over time. Time is a rather abstract or slippery concept in this context, but essentially we want to be able to change a network and observe those changes. A good place to start is to define a function that updates the network structure, in this case by forming or dissolving a single tie between two nodes.

Before the formal function is written, we can manually work through the process by which we would like to form a new tie or dissolve an existing tie. Starting with removing a tie is slightly easier. First, we need a way to identify all the existing ties for a particular node. In igraph there are a couple of equivalent ways to extract the adjacency list for a node in a network, which is a list of its direct ties. Here are the nodes that are tied to Node 24 in netdum. (To save some syntax space and protect us against inadvertent changes to the original data, the network is copied first.)

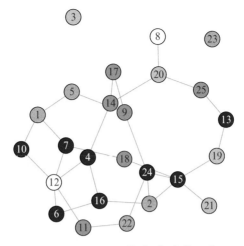

Fig. 13.1 Random test network with five levels of behavior indicated

```
g <- netdum
get.adjlist(g)[24]

  ## [[1]]
  ## + 4/25 vertices:
  ## [1]   2   9 15 22

g[[24,]]

  ## [[1]]
  ## + 4/25 vertices:
  ## [1]   2   9 15 22
```

Referring back to Fig. 13.1, you can see that Node 24 is connected to four other nodes (2, 9, 15, and 22). The get.adjlist() syntax is easier to decipher than the double-bracket shortcut, so that will be used hereafter.

In our model, we will want changes in ties to be driven by our node characteristic (Bh). Specifically, a reasonable model might suppose that friendship ties are more likely to be dissolved when the two friends are more dissimilar to each other on the behavior of interest. To model this we want to be able to compare the behavioral level of a particular network member with the behaviors of all of her friends. This is easy to do, building on the previous syntax. This uses the igraph vertex extractor function V(). Also, the get.adjlist() function returns a list, so the unlist() function is used to obtain a simple numeric vector. Finally, the stored adjacency list is used to filter the network to only display the behavior values for the nodes adjacent to Node 24. It looks like Node 24 has low amounts of exercise, which is similar to the level of Node 15. Node 24 is most dissimilar to Node 9, who has a very high level of exercise.

```
V(g)[24]$Bh
```

```
##  [1] 0.192
```

```
V_adj <- unlist(get.adjlist(g)[24])
V(g)[V_adj]$Bh
```

```
##  [1] 0.389 0.952 0.114 0.325
```

All of this can be combined into a simple dissimilarity vector that measures the absolute values of the differences between the Bh values of Node 24 and its adjacent nodes.

```
BhDiff <- abs(V(g)[V_adj]$Bh - V(g)[24]$Bh)
BhDiff
```

```
##  [1] 0.1966 0.7604 0.0783 0.1327
```

The next step is to use this information to select a tie that should be removed from the network. This can be done manually by identifying the two nodes and assigning FALSE to the node pair (again, by making a copy first):

```
gdum <- g
gdum[24,9] <- FALSE
get.adjlist(gdum)[24]
```

```
##  [[1]]
##  + 3/25 vertices:
##  [1]   2 15 22
```

A more programmatic way to select the tie to remove is to identify the most dissimilar pair of nodes. This can be done using index filtering (Fig. 13.2).

```
gdum <- g
V_sel <- V_adj[BhDiff == max(BhDiff)]
gdum[24,V_sel] <- FALSE
get.adjlist(gdum)[24]
```

```
##  [[1]]
##  + 3/25 vertices:
##  [1]   2 15 22
```

```
plot(gdum, layout = crd_save)
```

However, in our model we won't always want to remove the tie from most dissimilar pair of nodes. Instead, we would like to randomly remove ties where the

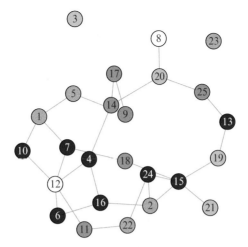

Fig. 13.2 Test network with one tie (24-9) removed

probability is weighted by the amount of dissimilarity. This adds some randomness and heterogeneity to the model. Once we have the dissimilarity vector, it can be used to do this type of random tie selection.

```
gdum <- g
V_sel <- sample(V_adj,1,prob=BhDiff)
gdum[24,V_sel] <- FALSE
get.adjlist(gdum)[24]

  ## [[1]]
  ## + 3/25 vertices:
  ## [1]   2 15 22
```

The sampling function selects one node from the list of adjacent nodes, with probability weighted by the dissimilarity vector. So, the more dissimilar the node pair is, the more likely it will be selected to be removed. To check that the sampling is working properly, we can sample multiple times and look at the selection distribution. This shows that Node 9 is selected most often, and Node 15 least often, which matches expectations based on the similarity on Bh values.

```
smplCheck <- sample(V_adj,500,replace=TRUE,prob=BhDiff)
table(smplCheck)

  ## smplCheck
  ##   2   9  15  22
  ##  91 327  35  47
```

Adding a new tie from a particular node to any other node proceeds in a similar fashion. The two main differences are that instead of picking the most dissimilar pair of nodes, we want to pick two nodes that are close to each other on the Bh characteristic. Second, because we are wanting to add a new tie, we need a way to select all of the non-adjacent nodes (i.e., nodes that are not directly tied to the target node).

The non-adjacent nodes can be selected by removing the adjacent nodes as well as the target vertex ID from a list of all the nodes. Nodes in igraph are numbered from 1 to the total size of the network. So, if we are still interested in Node 24, the following finds all non-adjacent nodes. This uses the vector indexing facility of R where values are dropped if a negative sign is used.

```
vtx <- 24
nodes <- 1:vcount(g)
V_nonadj <- nodes[-c(vtx,V_adj)]
V_nonadj

## [1]  1  3  4  5  6  7  8 10 11 12 13 14 16 17 18
## [16] 19 20 21 23 25
```

Following the same logic as before, we can now randomly create a new tie, based on the similarity between a pair of nodes. The inverse of the absolute differences is calculated, so now the vector BhDiff2 contains similarity scores.

```
BhDiff2 <- 1-abs(V(g)[V_nonadj]$Bh - V(g)[vtx]$Bh)
BhDiff2

## [1] 0.912 0.546 0.956 0.771 0.906 0.967 0.712
## [8] 0.900 0.207 0.663 0.859 0.309 0.997 0.194
## [15] 0.305 0.480 0.433 0.486 0.820 0.249

Sel_V <- sample(V_nonadj,1,prob=BhDiff2)
gnew <- g
gnew[vtx,Sel_V] <- TRUE
get.adjlist(gnew)[vtx]

## [[1]]
## + 5/25 vertices:
## [1]  2  9 10 15 22
```

The following code assigns a different color ("darkred") to the newly added tie, so that it can be seen easier in the plot (Fig. 13.3).

```
E(gnew)$color <- "grey"
E(gnew, P = c(vtx, Sel_V))$color <- "darkred"
plot(gnew, layout = crd_save)
```

All of this is preparation for creating a simple update function that can be called within a larger network simulation. The function that follows accepts a network (igraph) object and a target vertex. It first checks to see if the passed vertex in the network is an isolate, if it is then the function silently returns the unaltered network (because you can't remove a tie from an isolate). If the vertex is not an isolate, then the function randomly removes an existing tie with probability based on the dissimilarity between all tied pairs. It then adds a new tie with probability based on the similarity of all non-tied pairs. The two operations are combined into the same function so that the returned network object has the same density (i.e., number of total ties) as the original network. This makes the subsequent simulation easier to interpret. Note that this function relies on the igraph object already having the vertex attribute Bh defined.

```
Sel_update <- function(g,vtx){
  V_adj <- neighbors(g,vtx)
  if(length(V_adj)==0) return(g)
  BhDiff1 <- abs(V(g)[V_adj]$Bh - V(g)[vtx]$Bh)
  Sel_V <- sample(V_adj,1,prob=BhDiff1)
  g[vtx,Sel_V] <- FALSE
  nodes <- 1:vcount(g)
    V_nonadj <- nodes[-c(vtx,V_adj)]
  BhDiff2 <- 1-abs(V(g)[V_nonadj]$Bh - V(g)[vtx]$Bh)
  Sel_V <- sample(V_nonadj,1,prob=BhDiff2)
  g[vtx,Sel_V] <- TRUE
  g
}
```

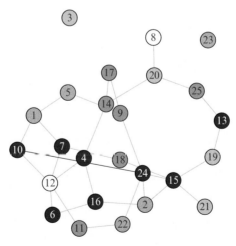

Fig. 13.3 Test network with one new tie added to Node 24

To test the function, pass it an igraph object along with the vertex whose ties are to be updated.

```
gtst <- g
node <- 24
gnew <- Sel_update(g,node)
neighbors(gtst,node)

  ## + 4/25 vertices:
  ## [1]  2  9 15 22

neighbors(gnew,node)

  ## + 4/25 vertices:
  ## [1]  2  7 15 22
```

13.1.1.3 Building a Simple Simulation of Social Selection

Now that an update function is available that randomly adds and drops ties in the abstract friendship network, a dynamic simulation model of social selection can be built. The following simulation model is simple, but has all of the elements of a dynamic network model. It operates on a network, makes changes over time, and those changes are observable.

In a dynamic model, time can be realistic or highly abstract. For the social selection model presented here, an abstract notion of time is used. Specifically, nodes will be selected randomly for updating, and we assume that there is some passage of time between each update. However, no further specific characteristics of time are provided or needed.

The following function Sel_sim encapsulates the social selection simulation. The function accepts an igraph network object and the number of desired updates. It starts by defining a list object that will be used to store the updated networks. Then inside a loop that runs for the number of desired updates, a random node is selected and the update function is called where an existing tie is removed, and a new tie added. The updated network is stored in the list object after each step, and after the loop is finished the entire network list is returned.

```
Sel_sim <- function(g,upd){
  g_lst <- lapply(1:(upd+1), function(i) i)
  g_lst[[1]] <- g
  for (i in 1:upd) {
    gnew <- g_lst[[i]]
    node <- sample(1:vcount(g),1)
    gupd <- Sel_update(gnew,node)
    g_lst[[i+1]] <- gupd
  }
  g_lst
}
```

This simulation function is inefficient in a few ways. First, the loop could possibly be replaced with a vectorized function. This could speed up the function, with some loss in readability. More importantly, the function stores the entire network for each update step. This will result in very large objects being returned by the function, based on the size of the network and number of updates. It is helpful at early stages of simulation development to preserve the whole networks, so that they can be examined. However, in later stages it would be typical to change the function to only return specific information about the networks, rather than the networks themselves.

The next step is to create a larger random network that will be used as input for the social selection simulation.

```
N <- 100
netdum <- erdos.renyi.game(N, p=0.10)
graph.density(netdum)

  ## [1] 0.0949

mean(degree(netdum))

  ## [1] 9.4

Bh <- runif(N,0,1)
BhCat <- cut(Bh, breaks=5, labels = FALSE)
V(netdum)$Bh <- Bh
V(netdum)$BhCat <- BhCat
table(V(netdum)$BhCat)

  ##
  ##  1  2  3  4  5
  ## 13 19 32 13 23
```

Now that a starting network has been created, the actual simulation is run and the results stored in a list of network objects. In this case, the simulation starts with the netdum network object, and it is run for 500 updates. The returned list of network objects contains the original network in the first position, and then 500 additional networks which correspond to each update in the simulation.

```
set.seed(999)
g_lst <- Sel_sim(netdum,500)
length(g_lst)

  ## [1] 501

summary(g_lst[[1]])

  ## IGRAPH U--- 100 470 -- Erdos renyi (gnp) graph
  ## + attr: name (g/c), type (g/c), loops (g/l),
  ## | p (g/n), Bh (v/n), BhCat (v/n)
```

13.1.1.4 Interpreting the Results of the Simulation

The results of the simulation should always be examined first to determine that the simulation ran as expected, and then relevant characteristics of the networks can be studied to see what patterns have emerged from the simulation modeling. Then it can be determined if the results inform some research question or hypothesis about the network dynamics.

A simple methods check for this simulation is that if the simulation worked as intended, then we should see no changes in density over the simulated networks. However, we should not see the same patterns of direct ties for any particular node in the network. (Note that the tie patterns will only be different for a particular node if that node was selected to be updated in the simulation. This is one reason to make sure to run the simulation many more times than the size of the network, to help ensure that all or at least most of the nodes have been updated.)

```
graph.density(g_lst[[1]])

  ## [1] 0.0949

graph.density(g_lst[[501]])

  ## [1] 0.0949

neighbors(g_lst[[1]],1)

  ## + 5/100 vertices:
  ## [1] 16 33 51 65 72

neighbors(g_lst[[501]],1)

  ## + 3/100 vertices:
  ## [1]  4 65 66
```

After it has been determined that the simulation is running properly, then more substantive assessments can be done. A basic hypothesis for this simple example is that we would expect the network to become more homophilous over time. This is because the tie updates are partially driven by the similarities of the nodes on the abstract behavioral characteristic. In this case, network modularity is a useful metric (see Chap. 8 for more information about modularity).

```
modularity(g_lst[[1]],BhCat)

  ## [1] -0.0221

modularity(g_lst[[501]],BhCat)

  ## [1] 0.168
```

Here we can see that the modularity was lower in the starting network compared to the final updated network. The higher modularity at the end tells us that ties between nodes in the same BhCat category are relatively more likely than ties between different categories. That is, connected nodes are more similar to each other than when the simulation started, thus demonstrating homophily.

This argument is more convincing if we use much more of the data provided by the simulation. The following plots show that there is substantial variability of modularity across the steps of the simulation. More importantly, modularity increases over time until near the end of the simulation run (Fig. 13.4).

```
sim_stat <- unlist(lapply(g_lst, function(u)
                   modularity(u,V(u)$BhCat)))
op <- par(mfrow=(c(1,2)))
plot(density(sim_stat),main="",xlab="Modularity")
plot(0:500,sim_stat,type="l",
     xlab="Simulation Step",ylab="Modularity")
par(op)
```

13.1.2 Simulating Social Influence

This next example focuses on a second dynamic network process that can explain homophily in social networks – namely, social influence. Social influence is the process by which behaviors (or attitudes, opinions, etc.) of an individual are influenced by the behaviors of those other persons who are close to them in their social network. Here, we develop a simple model of this process where an abstract behavior (Bh) for a particular member of a social network is influenced by the average behaviors of all of those to whom the person is directly tied. This constitutes a simple model of peer social influence. To make this slightly more realistic (and interesting) we will also build into the model the concept of a tolerance region. Every member of the network has a tolerance range (Tl). If an adjacent member has a value of (Bh) that falls outside of the tolerance range, then that person's behavior will not influence them. So, for example, if we again think of Bh as a measure of the level of physical activity, then in the following model the levels of physical activity of a network member's friends will influence her own level of activity, but only when those friends levels of activity are somewhat close to her own.

13.1.2.1 Setting Up the Simulation

The model building process is similar to the previous example, so we can proceed with slightly less exposition. The first step is to set up an example network. The only difference here is that we add a new vertex characteristic, Tl, the tolerance range for each network member. To start with we assume that every member has the same tolerance range of 0.20. (Remember that because of the random nature of the simulation, your network values will not match what is presented below.)

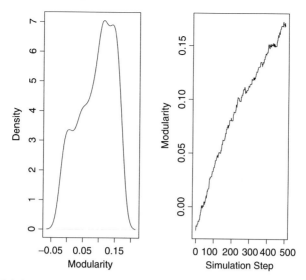

Fig. 13.4 Modularity over time in the social selection simulation

```
N <- 25
netdum <- erdos.renyi.game(N, p=0.10)
Bh <- runif(N,0,1)
BhCat <- cut(Bh, breaks=5, labels = FALSE)
V(netdum)$Bh <- Bh
V(netdum)$BhCat <- BhCat
V(netdum)$Tl <- 0.20
V(netdum)$Tl[1:10]
```

```
## [1] 0.2 0.2 0.2 0.2 0.2 0.2 0.2 0.2 0.2 0.2
```

13.1.2.2 Creating an Update Function

The first modeling step is once again to define a core update function. In this case, instead of updating ties, for social influence the vertex attribute Bh needs to be updated.

```
g <- netdum
V_adj <- neighbors(g,24)
V_adj
```

```
## + 4/25 vertices:
## [1] 2 9 15 22
```

```
V(g)[24]$Bh
```

```
## [1] 0.192
```

```
V(g)[V_adj]$Bh
```

```
## [1] 0.389 0.952 0.114 0.325
```

For Node 24 in the example network netdum, Bh starts quite low (0.19). The Bh values for Node 24's four neighbors range from 0.11 to 0.95.

A simple way to update the Bh value for Node 24 would be to take the mean of the starting value of Bh and the aggregate of the Bh values of all the neighbors. The new Bh value is 0.32, but this has been adjusted by the Bh values for all four neighbors. It has been particularly influenced by the high Bh value of Node 9 (0.95).

```
newval <-   .5*(V(g)$Bh[24]  + mean(V(g)[V_adj]$Bh))
newval
```

```
## [1] 0.318
```

The tolerance values can be used to filter out the nodes that have Bh values that fall outside of the tolerance range. So, after setting up vectors that store the Bh values, the neighbor Bh vector (N_Bh) is filtered here to only show the values for those neighbors whose Bh values are within 0.20 of Node 24's Bh value.

```
V_Bh <- V(g)[24]$Bh
V_Bh
```

```
## [1] 0.192
```

```
N_Bh <- V(g)[V_adj]$Bh
N_Bh
```

```
## [1] 0.389 0.952 0.114 0.325
```

```
N_Bh[abs(N_Bh-V_Bh) < .20]
```

```
## [1] 0.389 0.114 0.325
```

So now an updated Bh value for Node 24 can be calculated after removing the extremely high value for Node 9 that falls outside of the tolerance range. We see the updated Bh value for Node 24 has not changed quite as much as before.

```
newval2 <- .5*(V(g)$Bh[24] +
          mean(N_Bh[abs(N_Bh-V_Bh) < .20]))
newval2
```

```
## [1] 0.234
```

This filtering approach based on the absolute differences forms the heart of the following social influence update function. This function returns an updated Bh value for the selected vertex. If the selected vertex has no neighbors with Bh values inside her tolerance region, the Bh value is returned unchanged. Once again, this function assumes that the network object already has Bh and Tl vertex attributes.

```
Inf_update <- function(g,vtx){
   TL <- V(g)[vtx]$Tl
   V_adj <- neighbors(g,vtx)
   V_Bh <- V(g)[vtx]$Bh
   N_Bh <- V(g)[V_adj]$Bh
   ifelse(length(N_Bh[abs(N_Bh-V_Bh)<TL]) > 0,
          new_Bh <- .5*(V_Bh +
                    mean(N_Bh[abs(N_Bh-V_Bh) < TL])),
          new_Bh <- V_Bh
          )
   new_Bh
}
```

Testing out the new update function, it returns the correct value for Node 24.

```
newval3 <- Inf_update(g,24)
newval3

## [1] 0.234
```

13.1.2.3 Building the Simulation of Social Influence

Now that the social influence update function has been created, it can be put inside a function that runs the dynamic network model. This is similar to the social selection model in the previous section, with one big difference. It makes more sense to have every node in the network be influenced at the same time, because we assume that social influence is more of a continuous process. That means that instead of selecting nodes randomly to get updated, here we will update every node in the network for every run of the simulation. (Once again, the function presented here could be made more efficient in a number of ways.)

```
Inf_sim <- function(g,runs){

   g_lst <- lapply(1:(runs+1), function(i) i)
   g_lst[[1]] <- g
   for (i in 1:runs) {
      gnew <- g_lst[[i]]
```

```
    for (j in 1:length(V(g))) {
      V(gnew)[j]$Bh <- Inf_update(g=gnew,vtx=j)
    }

    g_lst[[i+1]] <- gnew
  }
  g_lst
}
```

The following code sets up the same type of random starting network, and then runs the influence model 50 times on the network. The simulation does not need to be run as long as the selection model, because every node gets updated for every run of the influence model.

```
N <- 100
netdum <- erdos.renyi.game(N, p=0.10)
Bh <- runif(N,0,1)
V(netdum)$Tl <- .20
V(netdum)$Bh <- Bh
V(netdum)$BhCat <- BhCat
set.seed(999)
g_lst <- Inf_sim(netdum,50)
```

13.1.2.4 Interpreting the Results of the Simulation

As the influence model runs, we might expect to see the variability of Bh to decrease because of the way that network members are adjusting their behaviors based on the average of their selected neighbors' behaviors (Fig. 13.5).

```
op <- par(mfrow=(c(3,2)))
plot(density(V(g_lst[[1]])$Bh),xlim=c(-.2,1.2),
     main="Original network")
plot(density(V(g_lst[[6]])$Bh),xlim-c(-.2,1.2),
     main='After 5 runs')
plot(density(V(g_lst[[11]])$Bh),xlim=c(-.2,1.2),
     main='After 10 runs')
plot(density(V(g_lst[[16]])$Bh),xlim=c(-.2,1.2),
     main='After 15 runs')
plot(density(V(g_lst[[26]])$Bh),xlim=c(-.2,1.2),
     main='After 25 runs')
plot(density(V(g_lst[[51]])$Bh),xlim=c(-.2,1.2),
     main='After 50 runs')
par(op)
```

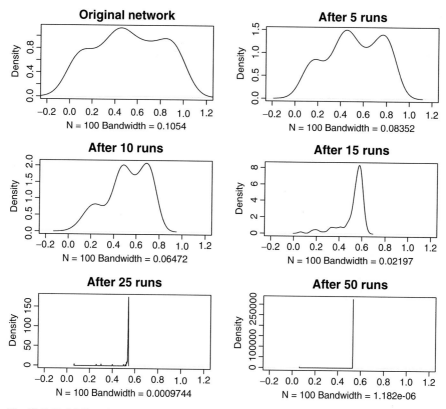

Fig. 13.5 Variability of Bh over time

The plots show us a few interesting things. First, it takes a few runs of the model to start shifting the behavioral variability. However, the homogenization process does not require 50 runs, by Run 25 almost all of the nodes have shifted their behavior to match those in the middle.

We can also see, however, that the network has become more homophilous. Figure 13.6 shows that by Run 25 most of the nodes fall into the middle Bh category, whereas they started out evenly distributed among the five categories.

```
V(g_lst[[1]])$BhCat <- cut(V(g_lst[[1]])$Bh,
          breaks=c(0,.2,.4,.6,.8,1), labels = FALSE)
V(g_lst[[26]])$BhCat <- cut(V(g_lst[[26]])$Bh,
          breaks=c(0,.2,.4,.6,.8,1), labels = FALSE)
V(g_lst[[51]])$BhCat <- cut(V(g_lst[[51]])$Bh,
          breaks=c(0,.2,.4,.6,.8,1), labels = FALSE)
```

```
V(g_lst[[1]])$color <- my_pal[V(g_lst[[1]])$BhCat]
V(g_lst[[26]])$color <- my_pal[V(g_lst[[26]])$BhCat]
op <- par(mfrow=c(1,2),mar=c(0,0,2,0))
plot(g_lst[[1]],vertex.label=NA,
     main="Original network")
plot(g_lst[[26]],vertex.label=NA,
     main="Network after Run 25")
par(op)
```

These simulation examples are intended to provide simple examples to illustrate
the power of R for exploring network and behavioral dynamics. They can be ext-
ended in a number of ways, both to learn how to build such simulations, but also to
apply them to more serious scientific questions. There are at least two simple ways
to extend the examples presented here. First, in the social influence simulation instead of having every network member start with the same tolerance range (0.20),
the effects of heterogeneous tolerance values on characteristics of social influence
could be explored. (Or, you can explore the effects of decreasing or increasing the
tolerance levels.) Second, the examples presented here separate out social influence
and social selection processes. The two simulations could be combined to explore
the characteristics of simultaneous selection and influence processes. (This is examined in Chap. 12 from a statistical modeling perspective.)

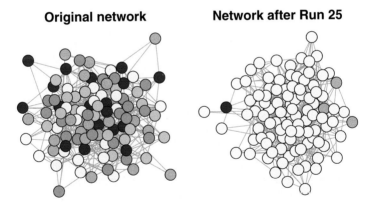

Fig. 13.6 Greater homophily over time

References

Barabási AL (2007) Network medicine from obesity to the diseasome. N Engl J Med. http://www.nejm.org/doi/full/10.1056/nejme078114

Barabási AL, Albert R (1999) Emergence of scaling in random networks. Science. http://www.sciencemag.org/content/286/5439/509.short

Broder A, Kumar R, Maghoul F (2000) Graph structure in the web. Computer Netw. http://www.sciencedirect.com/science/article/pii/S1389128600000839

Butts CT (2008) network: a package for managing relational data in R. J Stat Softw. http://cran.repo.bppt.go.id/web/packages/network/vignettes/networkVignette.pdf

Freeman LC (2004) The development of social network analysis: a study in the sociology of science. Empirical Press, p 205. ISBN:1594577145. https://books.google.com/books?id=VcxqQgAACAAJ&pgis=1

Fruchterman TMJ, Reingold EM (1991) Graph drawing by force-directed placement. Softw Pract Exp. ftp://132.180.22.143/axel/papers/reingold:graph_drawing_by_force_directed_placement.pdf

Galaskiewicz J (1985) The influence of corporate power, social status, and market position on corporate interlocks in a regional network. Social Forces. http://sf.oxfordjournals.org/content/64/2/403.abstract

Gentleman R, Lang DT (2007) Statistical analyses and reproducible research. J Comput Graph Stat. http://www.jstor.org/stable/27594227

Goodreau SM (2007) Advances in exponential random graph (p*) models applied to a large social network. Soc Netw 29(2):231–248. ISSN:03788733. doi:10.1016/j.socnet.2006.08.001

Granovetter MS (1973) The strength of weak ties. Am J Sociol. http://www.jstor.org/stable/2776392

Handcock MS et al (2008) statnet: software tools for the representation, visualization, analysis and simulation of network data. J Stat Softw 24(1):1–11. ISSN:1548-7660

© Springer International Publishing Switzerland 2015

D.A. Luke, *A User's Guide to Network Analysis in R*, Use R!,

DOI 10.1007/978-3-319-23883-8

Harris JK (2013) An introduction to exponential random graph modeling. SAGE, p 136. ISBN:148332205X. https://books.google.com/books?hl=en&lr=&id=lkYXBAAAQBAJ&pgis=1

Harris JK, Luke DA (2009) Forty years of secondhand smoke research: the gap between discovery and delivery. Am J Prev Med. http://www.sciencedirect.com/science/article/pii/S0749379709001548

Holland PW, Leinhardt S (1978) An omnibus test for social structure using triads. Sociol Methods Res. http://smr.sagepub.com/content/7/2/227.short

Hunter DR et al (2008) ergm: a package to fit, simulate and diagnose exponential-family models for networks. J Stat Softw 24(3):nihpa54860. ISSN:1548-7660.   http://www.pubmedcentral.nih.gov/articlerender.fcgi?artid=2743438&tool=pmcentrez&rendertype=abstract

Knoke D, Burt RS (1983) Prominence. Appl Netw Anal. https://scholar.google.com/scholar?q=knoke+burt+prominence&btnG=&hl=en&as_sdt=0%2C26#0

Kolaczyk ED (2009) Statistical analysis of network data: methods and models. Springer, p 398. ISBN:0387881468. https://books.google.com/books?id=Q-GNLsqq7QwC&pgis=1

Krebs VE (2002) Uncloaking terrorist networks. http://firstmonday.org/ojs/index.php/fm/article/view/941/863

Leischow SJ, Luke DA, et al. (2010) Mapping US government tobacco control leadership: networked for success? Nicotine Tob Res. http://ntr.oxfordjournals.org/content/12/9/888.short

Liljeros F, Edling CR, Amaral LAN (2001) The web of human sexual contacts. Nature. http://www.nature.com/nature/journal/v411/n6840/full/411907a0.html

Luke DA, Harris JK (2007) Network analysis in public health: history, methods, and applications. Annu Rev Public Health 28:69–93. ISSN:0163-7525. doi:10.1146/annurev.publhealth.28.021406.144132

Luke DA, Stamatakis KA (2012) Systems science methods in public health: dynamics, networks, and agents. Annu Rev Public Health. http://www.ncbi.nlm.nih.gov/pmc/articles/PMC3644212/

Luke DA, Wald LM (2013) Network influences on dissemination of evidence-based guidelines in state tobacco control programs. Health Educ Behav. http://heb.sagepub.com/content/40/1_suppl/33S.short

Luke DA, Harris JK, Shelton S (2010) Systems analysis of collaboration in 5 national tobacco control networks. Am J Public Health. http://www.ncbi.nlm.nih.gov/pmc/articles/PMC2882404/

McPherson M, Smith-Lovin L, Cook JM (2001) Birds of a feather: homophily in social networks. Annu Rev Sociol. http://www.jstor.org/stable/2678628

Morris M, Handcock MS, Hunter DR (2008) Specification of exponential-family random graph models: terms and computational aspects. J Stat Softw 24(4):1548–7660. ISSN:1548–7660. http://www.pubmedcentral.nih.gov/articlerender.fcgi?artid=2481518&tool=pmcentrez&rendertype=abstract

Murrell P (2005) R graphics. Taylor & Francis, p 328. ISBN:158488486X. https://books.google.com/books?id=fUUVngEACAAJ&pgis=1

Newman MEJ (2006) Modularity and community structure in networks. Proc Natl Acad Sci USA. http://www.pnas.org/content/103/23/8577.short

Newman M (2010) Networks: an introduction. Oxford University Press, Oxford, p 784. ISBN:0191500704. https://books.google.com/books?id=LrFaU4XCsUoC&pgis=1

Newman MEJ, Girvan M (2004) Finding and evaluating community structure in networks. Phys Rev E Stat Nonlinear Soft Matter Phys 69(2):1–15. ISSN:1063651X. doi:10.1103/PhysRevE.69.026113. arXiv: 0308217 [cond-mat]

Newman M, Barabási A-L, Watts DJ (2006) The structure and dynamics of networks. Princeton University Press, p 582. ISBN:0691113572. https://books.google.com/books?id=0FNQ1LYKTMwC&pgis=1

Rogers EM (2003) Diffusion of innovations, 5th edn. Simon and Schuster, p 576. ISBN:0743258231. https://books.google.com/books?id=9U1K5LjUOwEC&pgis=1

Scott J (2012) Social network analysis (3rd Ed.) SAGE Publications.

Scott J, Carrington PJ (2011) The SAGE handbook of social network analysis. SAGE Publications.

Snijders TAB, Pattison PE (2006) New specifications for exponential random graph models. Sociol Methodol. url: http://smx.sagepub.com/content/36/1/99.short.

Snijders TAB, Van de Bunt GG, Steglich CEG (2010) Introduction to stochastic actor-based models for network dynamics. Soc Netw. http://www.sciencedirect.com/science/article/pii/S0378873309000069

de Solla Price DJ (1976) A general theory of bibliometric and other cumulative advantage process. J Am Soc Info Sci. https://scholar.google.com/scholar?q=price+1976+bibliometric&btnG=&hl=en&as_sdt=0%2C26#4

Sporns O (2012) Discovering the human connectome. MIT, p 232. ISBN:026017903. https://books.google.com/books?id=uoNf2x0J8LMC&pgis=1

Tufte ER (1990) Envisioning information, vol 914. Graphics Press, p 126. https://books.google.com/books?id=1uloAAAAIAAJ&pgis=1

Tufte ER (2001) The visual display of quantitative information. Graphics Press, p 197. ISBN:0961392142. https://books.google.com/books?id=GTd5oQEACAAJ&pgis=1

Tukey JW (1977) Exploratory data analysis. Addison-Wesley, p 688. ISBN:0201076160. https://books.google.com/books?id=UT9dAAAAIAAJ&pgis=1

Valente TW (2010) Social networks and health: models, methods, and applications. Oxford University Press, Oxford, p 296. ISBN:0199719721. https://books.google.com/books?id=xnMzd1-7iGgC&pgis=1

Wasserman S, Faust K (1994) Social network analysis: methods and applications, vol 25. Cambridge University Press, p 825. ISBN:0521387078. https://books.google.com/books?id=CAm2DpIqRUIC&pgis=1

Watts DJ, Strogatz SH (1998) Collective dynamics of 'small-world' networks. Nature. http://www.nature.com/nature/journal/v393/n6684/abs/393440a0.html

Printed in the United States
By Bookmasters